McDonald's

Happy Meal® Toys

Around the World

1995-Present

Joyce and Terry Losonsky

4880 Lower Valley Road, Atglen, PA 19310 USA

Copyright © 1999 by Joyce and Terry Losonsky
Library of Congress Catalog Card Number: 99-65461

All rights reserved. No part of this work may be reproduced or used in any form or by any means—graphic, electronic, or mechanical, including photocopying or information storage and retrieval systems—without written permission from the copyright holder.
"Schiffer," "Schiffer Publishing Ltd. & Design," and the "Design of pen and ink well" are registered trademarks of Schiffer Publishing Ltd.

Designed by Bonnie M. Hensley
Type set in Humanist521 BT

ISBN: 0-7643-0960-9
Printed in China
1 2 3 4

Published by Schiffer Publishing Ltd.
4880 Lower Valley Road
Atglen, PA 19310
Phone: (610) 593-1777; Fax: (610) 593-2002
E-mail: Schifferbk@aol.com
Please visit our website catalog at **www.schifferbooks.com**

In Europe, Schiffer books are distributed by Bushwood Books
6 Marksbury Avenue Kew Gardens
Surrey TW9 4JF England
Phone: 44 (0)181 392-8585; Fax: 44 (0)181 392-9876
E-mail: Bushwd@aol.com

This book may be purchased from the publisher.
Include $3.95 for shipping. Please try your bookstore first.
We are interested in hearing from authors with book ideas on related subjects.
You may write for a free printed catalog.

Acknowledgments & Thanks

We would like to express our sincere appreciation to our friends for their endearing support. This book and our other books are written especially for collectors who have fun recording their latest treasure! We sincerely apologize for any names and familiar faces left off our list. A "Special Thanks and Hug" are extended to the following individuals:

Ken Clee
Ron & Eileen Corbett
Jimmy & Pat Futch
Bill & Pat Poe
Pat Sentell
Rich & Laurel Seidelman
E. J. Ritter
Kendall Mesker
Andrea & Milton Dodge
Natasha Losonsky
Nicole & Chuck Matlach
Ryan Losonsky
John & Eleanor Larsen
David Tuttle
Jan Antoine
Gerry Buchholtz
Luc Delaney - Canada
Nathalie & Jean Claude Royer - France
Kees & Conny Versteeg - Holland
Daphne Veerendad - Holland
Sonja Vos - Holland

Robin Murray - New Zealand
David Thornton - New Zealand
Julie Lush - New Zealand
Phil Hayes - New Zealand
Nigel Thomas - England
Gordon & Kathleen Fairgrieve - UK
Malcolm & Sheila Edwards - UK
Martin Waters - UK
Taylor & Cindy Wagon - Hawaii
Frank Duessel - Germany
Jurgen Seifried - Germany
Frank Schneidewind - Germany
Peter Peterson - Austria
Keith Hodges - Australia
Lexie Keady - Ireland
Irv & Robin Kirstein - Canada
Joyce Klassen - Canada
Don Wilson - Canada
Dave Archer - Canada
Gary Killops & family - Canada
David & Patrine Tang - Singapore
Varinda Rojanaumphai - Thailand
Larry & Manuella Poli - France
Camille Boone - USA
Glauco Carvalho - Brazil
Mabel Lo Mei Po - Hong Kong
Franck Lacaille - Guadeloupe

Contents

Chronology:
McDonald's at 45, Still Green and Growing (1955-2000)

Welcome to the world of McDonald's Happy Meal Toys Around the World! From humble beginnings, the late Ray Kroc opened his first McDonald's Drive-thru Restaurant in Des Plaines, Illinois (USA) on April 15, 1955. That day, rain and all, launched the sites and sounds of a chain of restaurants heard around the world. From downtown Tokyo to the Bavarian Alps in Germany, McDonald's restaurants around the world strive to offer the same quality food, quick service, clean surroundings as well as the value that customers have enjoyed for the last 45 years (1955-2000).

Ray Kroc, the founder of McDonald's Restaurants, had a saying: "When you're green, you're growing..." As the world's leading fast food restaurant organization heads toward its 45th birthday on April 15, 2000, McDonald's couldn't be greener! The McDonald's international restaurant story is not a simple business venture; it is one with many innovative twists and much marketing genius. It is the spirit and the people—the employees, customers, and the newest customers: the collectors—who enhance the operation. McDonald's story is the story of people helping people, employees helping customers, continually striving for customer satisfaction. It is the story of how the world's biggest small business continues to grow and prosper.

This book covers the period 1995 through mid 1999. We hope you share our feeling that McDonald's restaurants and McDonaldland—wherever located—are where the search for collecting fun and adventure truly begins!

1948 - Dick and Mac McDonald, known as the McDonald brothers, open their first limited menu, self-service McDonald's drive-in restaurant in San Bernardino, California (USA). Ray Kroc, a sales-

man for milk-shake machines, strikes an agreement with the McDonald brothers to franchise the concept of their restaurant. The menu consists of 15 cent hamburgers, 10 cent french fries, and 20 cent shakes. The restaurant is called "McDonald's" and the global chapter on the History of Fast Food Restaurants is beginning to unfold.

1952 - Dick and Mac McDonald officially franchise their McDonald's Speedee Service System. The initial building design was red and white tile with yellow neon arches going through the roof. The Golden Arches were born. The first eight McDonald's restaurants were located in southern California with a little hamburger man called "Speedee" as their mascot.

1955 - Ray Kroc opens his first restaurant in Des Plaines, Illinois (USA) on April 15, 1955. It is the ninth McDonald's to open in the USA, the other eight being in the southern California area. April 15th of each year is now officially known as Founder's Day. "Speedee" was the company symbol. By the end of that year, Ray Kroc had opened a second McDonald's in Fresno, California (USA).

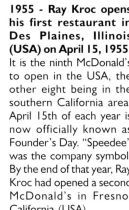

1956 - A dozen more McDonald's restaurants are added in Chicago, Illinois, and California (USA). Fred Turner, a very essential person, is hired to be the one man Operations Department. Mr. Turner initially worked as a grillman at the Des Plaines store.

1957 - Q.S.C. concept is instituted: Quality, Service and Cleanliness.

1959 - The 100th McDonald's restaurant is opened near Chicago's Midway Airport (USA).

1950s style USA McDonald's.

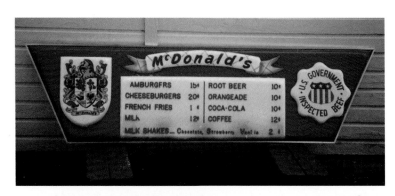

1950s USA menu board.

Speedee, the company symbol from 1955-1962.

"None of us is as good as all of us." – *Ray Kroc*

Ray Kroc (1903-1984), founder, McDonald's Corporation.

Two 1950s ads for McDonald's.

1960 - The 200th McDonald's restaurant is opened and "Look for the Golden Arches" jingle is played on radio. The first company owned and operated restaurant is opened in Columbus, Ohio (USA).

1961 - The All American Meal is launched for 45 cents and Ray Kroc buys out the McDonald brothers for $2.7 million. Hamburger University is opened in Elk Grove Village, Illinois. The first graduating class receives a degree in "Bachelor Hamburgerology." The management training center strove to standardize the product, place, price, and promotions along with the most important element: the selling agents, the employees. The All American Menu featured hamburgers, fries, and shakes. The bisected arches are introduced, an "M" slashed with a line symbolizes the Golden Arches and restaurant roof design. BOG - Be Our Guest cards are also introduced. Economist and marketing experts attempting to explain the secret of McDonald's early success agree that it was simply the controlled, meticulous way in which the company was operated that led to the success.

1962 - The Golden Arches replace Speedee as the McDonald's company symbol on packaging in January. "Go For Goodness at McDonald's" advertising slogan appears as McDonald's advertises nationally in *LIFE* magazine (USA).

1963 - Willard Scott appears as the first local Ronald McDonald, making his debut in Washington, D.C. at the Cherry Blossom Parade. The Filet-O-Fish sandwich is added to the menu at 24 cents. National advertisement appears in Reader's Digest magazine. The advertising blitz begins. Double hamburger and double cheeseburger are introduced. Ronald's Flying Hamburger is tested in Rockville, Maryland restaurant.

1964 - Archy McDonald logo is used on a limited number of premiums and bags. Arch Madden of Des Moines, Illinois turns the one day receipts of his six stores over to the Children's Zoo.

1965 - 10th Anniversary of McDonald's restaurants in the USA. McDonald's is on its way to becoming a "way of life." The building blocks for the coming years—training, research, advertising—are being anchored firmly into place.

1966 - Ronald McDonald makes his first national television appearance as the spokesman for McDonald's. McDonald's Corporation is listed on the New York Stock Exchange. In September of 1966, Ronald McDonald and his Flying Hamburger are introduced nationally as the new spokesman for McDonald's. Advertising jingle: "McDonald's—Where Quality Starts Fresh Everyday. Look for the Golden Arches." McDonald's is entrenched as a national name. Rather than concentrating on individual menu items, advertising focuses on creating a total quality image.

1967 - Canada (June 1) and Puerto Rico (November 10) opens their first McDonald's restaurants. British Columbia (Canada) was the first international restaurant for McDonald's outside the USA. OPNAD (The Operators' National Advertising Fund) is formed; it was the voluntary cooperative of McDonald's restaurants that purchased national advertising. Cooperative purchasing of products is instituted; cooperative warehousing is also developed among the stores in order to solve the storage problems created by increasing volume. McDonald's All-American High School band debuts in the Macy's Thanksgiving Day Parade in New York City. Ad theme, "McDonald's is Your Kind of Place." is introduced. Price of hamburgers increases from 15 cents to 18 cents on January 1, 1967. The roast beef sandwich is tested. Visitors to the Toronto Zoo in Canada enjoy McDonald's food in snack bars designed with natural wood to blend in with the environment. In 1993, McDonald's of Canada supports the NHL/Upper Deck trading card promotion which involves picturing the first female goalie ever to play NHL pre-season hockey, Manon Rheaume.

1968 - Hawaii opens its first McDonald's. The Big Mac sandwich and Hot Apple Pie are introduced. "Give up a pack for a Big Mac" theme slowly begins to take hold in the USA.

1969 - Red and white tile design is replaced with a "New Modern Design - The Mansard Roof." Changing designs with the time becomes a McDonald's standard practice. "QSC" is joined with "V" for value. Ronald McDonald receives a new costume design.

1960s era McDonald's cups.

1970 - **Virgin Islands (September 4)** and **Costa Rica (December 2) open their first McDonald's restaurants**. "You Deserve A Break Today" advertising slogan is born, along with McDonaldland. Christmas gift certificates are introduced. The price of a hamburger in the USA increases from 18 cents to 20 cents. With the opening of a McDonald's in Anchorage, Alaska, McDonald's is now in all fifty states of the USA.

1971 - **Guam (June 10), Japan (July 20), Holland (August 21), Panama (September 1), Germany (November 22), and Australia (December 20) open their first McDonald's restaurants**. Ronald McDonald's cast of characters begins to grow—Hamburglar, Grimace, Mayor McCheese, Captain Crook, and The Professor join Ronald McDonald in McDonaldland. McDonaldland is the imaginary place where the characters live and play, their playland. The first McDonald's Playland opens in Chula Vista, California (USA). The playlands play an important role in restaurants globally. "You Deserve A Break Today—So Get Up and Get Away to McDonald's" is the advertising theme. Cherry and Shamrock shakes are introduced. Quarter Pounder, Egg McMuffin, Tripple Ripple Ice cream, Fish 'n Chips, Cookies, Onion Rings, and Fried Chicken are in test sites in the USA. Some of the world's most beautiful McDonald's restaurants are located in international markets. The "oldest" McDonald's is located in Freiburg, West Germany, in a building dating back to 1250 A.D.

1972 - **France (June 30) and El Salvador (July 20) open McDonald's restaurants**. A total of 145 restaurants are now operating outside the USA with Canada having 96 restaurants operating in 1972. Ray Kroc receives the Horatio Alger Award and celebrates his 75th birthday. Large Fries are introduced. McDonald's participates in the Jerry Lewis Muscular Dystrophy Telethon for the first time. Danish is tested as a breakfast item. In the USA, McDonald's restaurants are either operated by independent business people or by the company. Expansion outside the USA is accomplished through the development of restaurants operated by (1) the company through subsidiaries; (2) franchisees—individuals granted franchises by the company, a subsidiary, or an affiliate; and (3) affiliates—companies where McDonald's equity is 50 percent or less and the remaining equity is generally owned by a resident national.

1973 - **Sweden (November 5) opens its first restaurant**. McDonald's makes the cover of *TIME* magazine and Canada's store #100 opens in St. John, New Brunswick. Hot Cakes, Sausage, and Soft Serve Cones are in testing stage. Egg McMuffin is added to the menu.

1974 - **Guatemala (June 19), Netherlands Antilles (August 16), England (October 1), and St. Thomas, Virgin Islands (November 25) open restaurants**. England adds the 3000th store in Woolwich, England. The first Ronald McDonald House opens in Philadelphia, Pennsylvania (USA). McDonaldland Cookies become a menu item. QLT/HLT, Iced Tea, Sundaes, and Diet Drinks are in testing stage. International McDonald's strives to strengthen local economies by supporting local charitable, civic, educational, and community service programs. This approach is in keeping with Ray Kroc's original philosophy of "putting something back into the communities where McDonald's does business."

1975 - **McDonald's Systems of Europe, Inc., based in Frankfurt, Germany is formed**. A Stockholm, Sweden restaurant is first to reach $2 million in annual sales. The Drive-thru concept is opened in Sierra Vista, Arizona and Oklahoma City, Oklahoma (USA). The Honorary Meal of McDonaldland is introduced in the USA and Canada. "Twoallbeefpattiesspecialsaucelettucecheesespicklesonionsonasesameseedbun" promotional jingle catches the imagination of the country. The direct relationship between advertising and sales is rooted. McDonald's celebrates its 20th Anniversary. Ad campaign, "We Do It All For You" is introduced. New polystyrene packaging begins. McFeast, Salad Bar, Scrambled Eggs, McChicken Sandwich, and Chili are in testing stage.

1976 - **New Zealand (June 7) and Switzerland (October 20) open their first restaurants**. Montreal, Canada opens McDonald's 4000th restaurant (4000 Ste-Catherin Quest, Montreal, Quebec on July 13, 1976). Kisarazu, Japan opens Japan's 100th restaurant. "You, You're The One" advertising campaign focuses on individual attention to detail, the customer. The breakfast menu is instituted. McDonald's becomes the official sponsor of Olympic Games in Montreal, Canada. New Zealand does not specifically distribute a Happy Meal premium for children under the age of 3 (USA) or the age of 5, as in Germany. In New Zealand, about 8-10 different Happy Meal boxes tend to appear and disappear with various promotions. These same box designs just kept rotating around in the different cities, rather than any specific distribution cycle. New Zealand provides "Lobby Toys" which compensate for the lack of U-3 toys.

1977 - **Ireland (May 9) and Austria (July 21) open McDonald's restaurants**. The Happy Meal concept is test marketed in the USA. From the Happy Plate, Happy Cup, and Happy Hat, the overall Happy Meal concept evolves. The initial boxes are designed like round top metal lunch boxes. McDonald's owns over half of all of its real estate sites by the end of the year; a key marketing decision for future success. Chocolaty Chip Cookies, Chopped Beef Sandwich, Sausage McMuffin, Chicken Pot Pie, and Onion Nuggets are in testing stage.

1978 - **Belgium (March 21) opens its first restaurant**. Osaka, Japan is the first to reach $3 million in sales. Kanagawa, Japan (Fujisawa City) opens store #5000 on October 17, 1978.

1979 - **Brazil (February 13) and Singapore (October 20) open McDonald's restaurants**. The 100th McDonald's in Australia, featuring a Drive-thru, opens in Sunnybank. The 100th German McDonald's opens in Hamburg. New advertising theme, "Nobody Can Do It Like McDonald's Can" hits the airwaves. Chicken McNuggets, Ham Biscuits, McCola, McPizza, and Cinnamon Streusel are in testing stage. Happy Meal tests are conducted in the USA while the advertising slogans focus on children and the "Collect all..." theme.

1970s Ronald McDonald doll, USA.

1970s McDonald's lighter.

1980 - 25th Anniversary (Silver) celebrated. Birdie the Early Bird joins McDonaldland and Ronald's cast of characters. Ronald McDonald Houses are providing shelter for 33,000 families in the USA, Canada, and Australia. The 1000th International McDonald's opens in Hong Kong. Store #6000 opens on June 23, 1980 in Munich, Germany. In 1994, McDonald's of Germany holds the first ever world press conference in Munich, Germany to announce outstanding financial results from the 500 restaurants in Germany. Newest advertising slogan, "You Deserve a Break Today, and Nobody Makes Your Day Like McDonald's Can," appears in the mass media.

1981 - Spain (March 10), Denmark (April 15), and the Philippines (September 27) open McDonald's restaurants. McDonald's becomes the largest food service organization in Canada. The first Ronald McDonald House opens in Toronto, Canada. Australia, Germany, Guam, Holland, Japan, and Panama celebrate their 10th Anniversary. Japan is involved in an extensive test market program for the Happy Meal promotions and Kid's give-away premiums. The items are tested and then run about nine months to a year later if they are successful in drawing an interest and increasing sales.

1982 - Malaysia (April 29) opens a McDonald's restaurant. Soft serve cones are added to the menu.

1983 - Norway (November 18) opens its first McDonald's. Yugoslavia signs joint venture to open a McDonald's restaurant. The 100th McDonald's in England opens in Manchester. McNugget Mania promotes Chicken McNuggets in addition to the menu. "McDonald's and You" advertising campaign is introduced. Diet Coke added to the menu; McDLT is in testing stage. Ronald McDonald learns Italian and communicates in at least eighteen languages. England typically distributes four different Happy Meal boxes with the Kid's Meal. They are designed to serve as a theme backdrop, with punched out pieces enhancing the 3-D effect of the packaging.

1984 - Taiwan (January 28), Andorra (June 29), Finland (December 14), and Wales (December 3) add their first restaurants. Ray Kroc dies at the age of 81, but his vision lives on. The newest ad campaign, "It's a Good Time for the Great Taste of McDonald's," refreshes the global thirst for McDonald's. At year's end there are seventy-three Ronald McDonald Houses operating in the USA, Canada, and Australia. McDonald's and its franchisees provide money to build the Olympic Swim Stadium in Los Angeles. "When the US Wins, You Win" promotion, in conjunction with the Summer Olympics, wins customers. Pecan roll and Bacon/Egg/Cheese Biscuit are in testing stage. There are 1,709 restaurants in 34 countries and territories outside the USA in 1984. In 1984, five previously opened international markets—Japan, Hong Kong, Sweden, Taiwan, and Switzerland—represent 20 percent of McDonald's systemwide sales. Ronald McDonald Children's Charities (RMCC) is founded in 1984 in memory of Ray Kroc. The cornerstone of RMCC is the Ronald McDonald House program. A Ronald McDonald House is a "home away from home" for out of town families of children receiving hospital treatment for life threatening illnesses such as cancer. The houses provide a supportive environment for parents and siblings of sick children. Introduced in Philadelphia, Pennsylvania (USA) in 1974, by 1984 there are more than 150 Ronald McDonald Houses worldwide and many more under development. The New Zealand Ronald McDonald House, located beside the Wellington Hospital, becomes McDonald's 151st house.. The first Ronald McDonald House opens in the former communist East Bloc in 1993. McDonald's of Finland, in conjunction with the Finnish Ministry for Internal Affairs, The Ministry of Transportation, and Finland's First Alert Company, (Torres), develop a program to inform residents of the new "112" emergency number, which functions like the "911" number in the USA.

1985 - Thailand (February 23), Luxembourg (July 17), Bermuda (July 24), Venezuela (August 31), Italy (October 15), Mexico (October 29), and Aruba (April 4) open restaurants and help McDonald's celebrate its 30th Anniversary. The Ronald McDonald House program offers a "home away from home" for families of children being treated for serious illnesses. The first European Ronald McDonald House opens in March, 1985 in Amsterdam. A new Braille menu assist visually impaired customers. Customers sing, "The Hot Stays Hot and the Cool Stays Cool" as McDonald's introduces the McDLT sandwich. Ad theme, "Good Time for Great Taste," encourages customers to visit their local McDonald's. Sausage McMuffin is added to the menu. The famous picture of the Buddhist priest eating french fries on the steps of an ancient Japanese temple (1985 McDonald's Calendar) amply illustrates the worldwide appeal of McDonald's french fries. Mexico calls their Happy Meal, "La Cajita Feliz."

1986 - Cuba (Navy Base-April 24), Turkey (October 24), and Argentina (November 24) open McDonald's restaurants. Store #9000 opens in Sydney, Australia. Three varieties of biscuit sandwiches and Sausage McMuffin with Egg are introduced. The Special McNuggets Shanghai promotion features chopsticks and McFortune cookies. Germany begins to call their children's menu *Kindermenu* in 1986 with their Ship Shape promotion.

1987 - Macau (April 11) and Scotland (November 23) open their first McDonald's. Salads become the newest menu item. The Monopoly game is the most successful national promotion and Muppet Babies is the most successful Happy Meal to date. Rome, Italy restaurant reaches $5 million in sales. Rock 'n Roll McDonald's in Chicago, Illinois (USA) becomes the first USA store to reach $5 million in sales. Mac Tonight debuts in promotional literature and ad campaigns. McDonald's of Canada celebrates its 20th Anniversary. Canada's boxes and bags for the "Joyuex Festin" Happy Meal are the same as the USA except that they are printed in English and French.

1988 - Yugoslavia (March 22), Korea (March 29), and Hungary (April 30) open restaurants. At year's end there are 168 Ronald McDonald Houses worldwide. The newest ad theme, "Good Time, Great Taste, That's Why This is My Place," appears in commercials. CosMc, the little space alien joins McDonaldland. New breakfast theme, "America's Morning Place," encourages the breakfast market. Cheddar Melt is added to menu. McMasters program to recruit older workers is introduced. Bonus size fries and super size soft drinks are standard menu items in the USA. German archives indicate January-April 1988 in Augsburg, Germany is where the Junior-Tüte (Happy Meal) was officially test marketed.

1989 - McChicken sandwich is introduced nationally in the USA after ten years of testing. McDonald's sells 35 million plush Muppet Baby Dolls to raise nearly $9 million for RMCC and Ronald McDonald Houses. Ronald McDonald Houses in the USA, Canada, Australia, Austria, Germany and the Netherlands serve over one million family members per year. Store #11,000 opens on October 20th in Hong Kong.

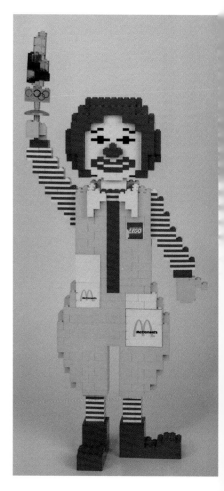

1980s USA Happy Meal display.

8

1990 - Russia (January 31), China-Shenzhen (October 8), and Chile (November 19) open restaurants to long lines of customers. Movement to recycled paper products reflects environmental concerns. The 35th Anniversary is celebrated. McDonald's food is served in two restaurant cars on the Swiss Federal RR Basel-Geneva run. Newest ad campaign,: "Food, Folks and Fun," is introduced. McDonald's opens in Shenzhen, a special economic zone of the People's Republic of China, just across the border from Hong Kong.

1991 - Indonesia (February 22), Portugal (May 23), Northern Ireland (October 14), Greece (November 12), Uruguay (November 18), and Martinique (December 16) are added. A total of 427 international restaurants are added in 1991. Germany begins issuing Junior-Tute (Happy Meal) in a brown bag with comics on the back. McDonald's international goal is to make the Golden Arches the customer's first choice around the globe.

1992 - Czech Republic (March 20), Guadeloupe (April 16), Poland (June 16), Monaco (November 20), Brunei (December), and Morocco (December 18) are opened, making a total of sixty-four countries in which McDonald's does business. The 25th Anniversary of McDonald's International is celebrated. McDonald's of Sweden develops educational maps/trayliners to illustrate the changes that occurred in Europe, including a picture of the thirteen "New Countries" that were created. The New Europe Map is distributed as a trayliner in September 1992. Food, Folks, and Fun along with a geography lesson are the offerings.

1993 - Saipan (March 18), Iceland (September 3), Israel (October 14), Slovenia (December 4), and Saudi Arabia (December 8) are added to the roster of new restaurants. McDonald's participates in the exhibition, "Design, Mirror of The Century" at the Grand Palais in Paris, France. An exact replica of the old red and white restaurant in Des Plaines, Illinois is constructed to serve as a fully operational restaurant at the Grand Palais. A special trayliner is issued, depicting the outstanding success of the operation. The Minister of Culture, Mr. Jacques Toubon, declares the exhibit, "*C'est Magnifique!*" McDonald's provides something more than food: a consistent quality approach, 38 years later. German Junior-Tute Happy Meal is changed to white numbered bags with comics on the back. McWorld advertising emphasizes the global relationship between McDonald's and the Earth.

1994 - Oman (March 30), Kuwait (June 15), New Caledonia (July 26), Egypt (October 20), Trinidad (November 12), Bulgaria (December 10), Bahrain (December 15), Latvia (December 15), and United Arab Emerates (December 21) are the newest places McDonald's calls home. The 500th McDonald's in Germany opens and the new McDonald's in Budapest becomes the 22nd in the capital of Hungary. It is the 100th restaurant in Central Europe. McDonald's announces planned operation of restaurants at the 1996 Olympic Village in Atlanta, Georgia (USA). McWorld Environmental Stamp Design Contest produces a series of four stamps in 1995. The 10th Anniversary of Ronald McDonald Children's Charities (RMCC) and the 20th anniversary of the first Ronald McDonald House are both celebrated. There are over 162 Ronald McDonald houses worldwide, including houses in Rio de Janeiro, Brazil, and Auckland, New Zealand. Satellite restaurants in Wal Mart and limited space locations continue to grow in number. There are over one hundred satellite locations in Canada, ten in Mexico, and two in Puerto Rico. These low-cost units serve a simplified menu. By year's end, China has over thirty restaurants and is growing rapidly to meet the population demands of thirty million people. In the Asia/Pacific region, McDonald's has only one restaurant for every 900,000 people, compared to one for every 25,000

people in the USA. Japan opens its 1,000th restaurant in 1994 and the United Kingdom and Germany are already at 500 restaurants and growing. They employ more then 650,000 people around the world. Ronald McDonald speaks well over forty languages. Menu selections vary slightly in adapting traditional McDonald's offerings to local tastes: special chicken dishes in Japan, veggie burgers in the Netherlands, or Kahuna burgers in Australia. Germany distributes Batman, The Animated Series Happy Meal, 1994 in a Happy Meal Box (four different boxes). This was the only Happy Meal in Germany distributed in a box due to ecological concerns. Boxes and toys are made in Germany and distributed in other European countries.

1995 - Romania (June 16), Malta (July 7), Columbia (July 14), Jamaica (September 28), Slovakia (October 13), South Africa (November 11), Jersey (December 1), Qatar (December 13), Honduras (December 14), and St. Martin (December 15) combine to make the total eighty-nine international countries and still growing! The goal is over one hundred countries by the year 1996. McDonald's has just begun to communicate the

1990s Happy Meal Workshop advertising.

McDonald's way in the USA, as well as communicate the process on an international scale. By 1995, over half of McDonald's income stream comes from outside the USA. The company expands into smaller sales units with McStop and McSnack operations in retail stores and limited space locations. McDonald's restaurants span the globe and are located on all continents except Antarctica. The name of the Japanese Happy Meal set is changed to "Happy Set" from "Okosame Set" and the first Happy Set is the Sonic Happy Set. McDonald's believes that preserving and enhancing the integrity of one's environment benefits everyone. As Ray Kroc stated, "None of us is as good as all of us." Ray Kroc reflected in 1965, "..I like to try to dream of what this Company of ours will be like on its 50th anniversary—the golden anniversary of the Golden Arches. I try to imagine but I admit I cannot. After all, who ten years ago could have predicted we would be where we are today..." Looking back over the last forty years Ray Kroc would be justifiably proud of the amazing record of growth and success. McDonald's is an institution, a way of life, and is world class. McDonald's is a global beacon around the world.

More 1990s Happy Meal Workshop advertising.

1996 - Croatia (February 2), Western Samoa (March 2), Fiji (May 1), Liechtenstein (May 3), Lithuania (May 31), India (October 13), Peru (October 17), Jordan (November 7), Paraguay (November 21), Dominican Republic (November 30), Belarus (December 10), and Tahiti (December 10) make a total of 101 countries with McDonald's restaurants by 1996. India's restaurants are the first McDonald's in the world without beef on the menu. Menu includes chicken, fish, potatoes, bread and soft drinks. Vegetables burgers are the hot item! The Ronald McDonald House Program and the Ronald McDonald Children's Charities combine to form one entity: Ronald McDonald House Charities.

1997 - Ukraine (May 24), Cyprus (June 12), Macedonia (September 6), Ecuador (October 9), Bolivia (October 24), Reunion Island (December 14), Isle of Man (December 15), and Surinam (December 18) are the smaller countries entered into in 1997. There is still no single coordinated program for the distribution of Happy Meal toys globally. Each country is different, with various countries joining together at times. Each Pacific Rim country is independent as are Central and South America. Eastern Europe, with the exception of Hungary and the Middle East, is not specifically involved in a coordinated program. The UK (United Kingdom) coordinates programs for England, Scotland, Wales, Ireland, and Iceland. Germany, France, Spain, Portugal, Italy, and Andorra are independent. Scandinavia, Switzerland, Belgium, and Holland are coordinated by the London based McDonald's Development Corporation. In essence, all of Europe may run the same program—like the World Cup Promotion—but have some different premiums with different packaging. The packaging tends to be multilingual with multilingual graphics or can be totally different for each country. Europe and Japan tend to run National and Test Programs as well as to rerun Happy Meal promotions from two to three years prior in the USA, reducing the number of toys from eight to four. The USA tends to distribute eight premiums on an average each month, the exception being Barbie/Hot Wheels, which have sixteen premiums offered per month.

1998 - Moldova (April 30), Nicaragua (July 11), Lebanon (September 18), Pakistan (September 19), and Sri Lanka (October 16) join the ranks of McDonald's global family. There is a kind of cross-pollination, with customers, collectors, and restaurants all over the world exchanging their specialties with each other—like the Australian meat pie being sold in England. The same thing is happening with collectors all over the world; they are exchanging their Happy Meal toys and promotional items with each other, such as a New Zealand Big Buddies traded for a set of USA Moveables. The McDonald's success story is one of recognition and respect for multicultural differences. It is more than the story of business, it is the story of people meeting people.

1999 - In response to the often asked question, "What is the earliest international Happy Meal?" (a Happy Meal being a hamburger, cheeseburger, fries, and drink with a toy inside a designated carton or bag) one must reply that the answer is difficult because of cultural differences among countries. Some countries, such as Canada, did not include the toy in the Happy Meal price, selling it separately instead. Notice the prices on the Canadian toys. In the USA, the toy is included with the price of the Happy Meal. At different times, the toy may be included with the meal and/or sold separately. Thus, the answer is not straightforward but cloudy. The earliest labeled "International Happy Meal" was the 1980 "Ronald McDonald Children's Meal" from Australia, the Caribbean, Hong Kong, Latin America, and New Zealand. This meal offered a hamburger or cheeseburger, fries, soft drink, and possibly a toy or premium in a "Happy Meal" style box. There were seven Happy Meal boxes with assorted generic premiums. The next recorded "Happy Meal" promotion was the "Airplane" meal, which ran in 1982. This promotion consisted of boxes with punch out wings to form airplanes from the boxes. The "toy" was the specially designed box, after it was perforated and assembled. Many countries have given out premiums that are not specifically associated with the words, "Happy Meal." Some of these promotions are called self-liquidators, which are tied to offers to purchase a particular product with the purchase of a food item or Value Meal. Prior to 1999, there was no coordinated program for International Happy Meal offerings. Coordinated efforts around the world began in 1999.

2000 - Moving towards the Year 2000 and the 21st Century, it appears that McDonald's is moving in the direction of coordinating large segments of the global population. As Y2K (Year 2000) countries team together, Canada, the USA, and Mexico are running a coordinated promotion using multilingual packaging for premiums.

Have a Happy Meal Morning!

NOW! McDonald's Breakfast Happy Meal!

Hot Cakes Happy Meal
Hot Cakes & Syrup
Juice or Lowfat Milk
TOY!
$1 99

1990s Breakfast Happy Meal display.

1990s Happy Meal toy set from France, 19" high.

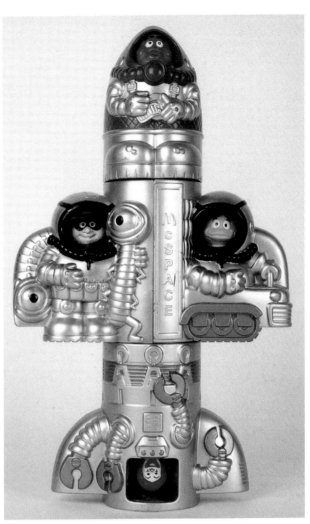

1990s Happy Meal toy set from New Zealand.

1990s McDonald's advertising
from Saudi Arabia.

1990s advertising from Saudi Arabia for McDonaldland Cookies.

1990s Happy Meal bag from France.

1990s trayliner from Singapore.

10th Happy Meal Anniversary in New Zealand.

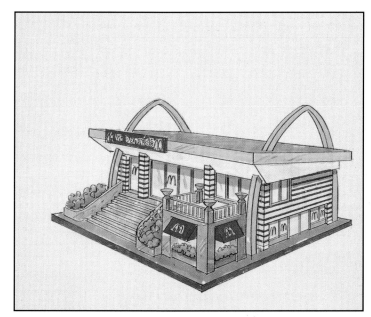

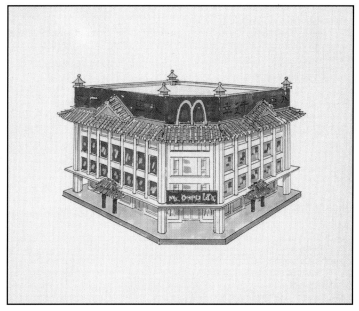

Design for 1990s McDonald's in Singapore.

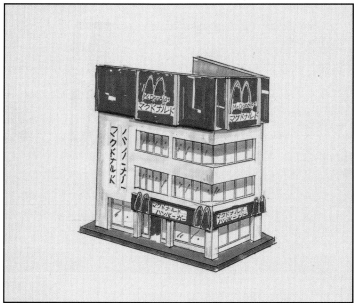

Designs for two different style 1990s McDonald's in Japan.

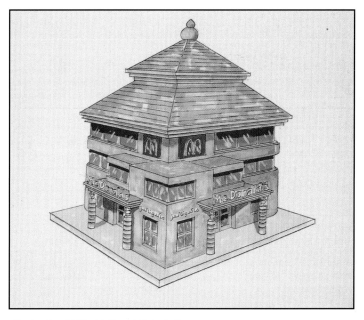

Design for 1990s McDonald's in India.

Helpful Information for Using this Book

Realizing that McDonald's is the largest and best-known global food service retailer, with more than 17,000 locations in 114 countries, the process of organizing the collectible items is both global and geometric. The following codes were used to identify the systemwide restaurants:

Country	Code	Opening Date	# in 1999
1. United States	USA	April 15, 1955	12,190
2. Canada	Can	June 1, 1967	1,063
3. Puerto Rico	Pue	November 10, 1967	116
4. Virgin Islands	Vir	September 4, 1970	6
5. Costa Rica	Cos	December 28, 1970	20
6. Guam	Gum	June 10, 1971	6
7. Japan	Jpn	July 20, 1971	2,640
8. Netherlands	Net	August 21, 1971	178
9. Panama	Pan	September 1, 1971	21
10. Germany	Ger	November 22, 1971	878
11. Australia	Aus	December 30, 1971	659
12. France	Fra	June 30, 1972	680
13. El Salvador	Sal	July 20, 1972	1
14. Sweden	Swe	November 5, 1973	166
15. Guatemala	Gua	June 19, 1974	24
16. Netherland Antilles	Nea	August 16, 1974	24
17. England	Uk	October 1, 1974	779
18. Hong Kong	Hon	January 8, 1995	144
19. Bahamas	Bah	August 4, 1975	3
20. New Zealand	Zea	June 7, 1976	144
21. Switzerland	Swi	October 20, 1976	94
22. Ireland	Ire	May 9, 1977	40
23. Austria	Ast	July 21, 1977	109
24. Belgium	Bel	March 21, 1978	62
25. Brazil	Bra	February 13, 1979	512
26. Singapore	Sin	October 20, 1979	104
27. Spain	Spa	March 10, 1981	163
28. Denmark	Den	April 15, 1981	82
29. Philippines	Phi	September 27, 1981	178
30. Malaysia	Mal	April 29, 1982	114
31. Norway	Nor	November 18, 1983	44
32. Taiwan	Tai	January 28, 1984	273
33. Andorra	And	June 29, 1984	2
34. Wales	Wal	December 3, 1984	31
35. Finland	Fin	December 14, 1984	85
36. Thailand	Tha	February 23, 1985	63
37. Aruba	Aru	April 4, 1985	2
38. Luxembourg	Lux	July 17, 1985	4
39. Venezuela	Ven	August 31, 1985	57
40. Italy	Ita	October 15, 1985	183
41. Mexico	Mex	October 29, 1985	139
42. Cuba	Cub	April 24, 1986	1
43. Turkey	Tur	October 24, 1986	94
44. Argentina	Arg	November 24, 1986	148
45. Macau	Mac	April 11, 1987	10
46. Scotland	Sco	November 23, 1987	47
47. Yugoslavia	Yug	March 22, 1988	14
48. Korea	Kor	March 29, 1988	127
49. Hungary	Hun	April 30, 1988	64
50. Russia	Rus	January 31, 1990	39
51. China	Chi	October 8, 1990	207
52. Chile	Chl	November 19, 1990	30
53. Indonesia	Idn	February 22, 1991	103
54. Portugal	Por	May 23, 1991	50
55. Northern Ireland	Nir	October 14, 1991	11
56. Greece	Gre	November 12, 1991	38
57. Uruguay	Uru	November 18, 1991	21
58. Martinique	Mar	December 16, 1991	5
59. Czech Republic	Cze	March 20, 1992	45
60. Guadeloupe	Gle	April 6, 1992	4
61. Poland	Pol	June 16, 1992	109
62. Monaco	Mon	November 20, 1992	1
63. Brunei	Bru	December 1, 1992	1
64. Morocco	Mor	December 18, 1992	6
65. Saipan	Sai	March 18, 1992	2
66. Iceland	Ice	September 3, 1993	2
67. Israel	Isr	October 14, 1993	57
68. Slovenia	Slo	December 4, 1993	11
69. Saudi Arabia	Sau	December 8, 1993	30
70. Oman	Oma	March 30, 1994	2
71. Kuwait	Kua	June 15, 1994	16
72. New Caledonia	Nec	July 26, 1994	1
73. Egypt	Egy	October 20, 1994	25
74. Trinidad	Tri	November 12, 1994	4
75. Bulgaria	Bul	December 10, 1994	11
76. Bahrain	Bah	December 15, 1994	5
77. Latvia	Lat	December 15, 1994	6
78. United Arab Emirates	Uar	December 21, 1994	13
79. Estonia	Est	April 29, 1995	5
80. Romania	Rom	June 16, 1995	32
81. Malta	Mal	July 7, 1995	6
82. Columbia	Col	July 14, 1995	18
83. Jamaica	Jam	September 28, 1995	9
84. Slovakia	Slv	October 13, 1995	18
85. South Africa	Saf	November 11, 1995	44
86. Jersey	Jer	December 1, 1995	
87. Qatar	Qat	December 13, 1995	3
88. Honduras	Hon	December 14, 1995	5
89. St. Martin	Stm	December 15, 1995	1
90. Croatia	Cro	February 2, 1996	10
91. Western Samoa	Wes	March 2, 1996	1
92. Fiji	Fij	May 1, 1996	2
93. Liechtenstein	Lie	May 3, 1996	1
94. Lithuania	Lit	May 31, 1996	5
95. India	Ind	October 13, 1996	11
96. Peru	Per	October 17, 1996	7
97. Jordan	Jor	November 7, 1996	3
98. Paraguay	Par	November 21, 1996	6
99. Dominican Republic	Dor	November 30, 1996	8
100. Belarus	Bel	December 10, 1996	5
101. Tahiti	Tah	December 10, 1996	1
102. Ukraine	Ukr	May 24, 1997	10
103. Cyprus	Cyp	June 12, 1997	4
104. Macedonia	Mac	September 6, 1997	1
105. Ecuador	Ecu	October 9, 1997	4
106. Bolivia	Bol	October 24, 1997	3
107. Reunion Island	Rei	December 14, 1997	2
108. Isle of Man	Ism	December 15, 1997	1
109. Surinam	Sur	December 18, 1997	1
110. Moldova	Mol	April 30, 1998	1
111. Nicaragua	Nic	July 11, 1998	1
112. Lebanon	Leb	September 18, 1998	1
113. Pakistan	Pak	September 19, 1998	1
114. Sri Lanka	Sri	October 16, 1998	1

PRICING - The price range listed is for MINT IN THE PACKAGE and/or MINT ON THE TREE. Loose toys are generally 50% less than the lowest mint in the package price listed. Damaged, chipped, or broken toys tend to have little value with a collector. The real value of any collectible is what a buyer is willing to pay for a particular item at a particular time. This value may exceed the stated mint in the package (MIP) or mint on tree (MIT) range. Likewise, since McDonald's makes millions of toys, value may be over inflated based on regional or global markets. Price ranges vary by geographical regions, continents, and availability, however the price range listed can be used as a relative guide for drawing comparisons among items.

Note that in most cases the dealer prices for these toys will be more, because dealers acquired their toys when prices were considerably higher. They are holding toys at higher prices because they believe the market will rebound. That is, dealers believe young people who have worked at McDonald's will come full circle and begin to collect memories from their youth, just as their parents and grandparents did in the earlier years.

Premium Names - The toys are listed by the names on the packaging whenever possible.

Check-Off Blocks - The check-off blocks are provided for record keeping. Mark one block for mint in package; one block for loose.

Box Names - The boxes were named by the authors with the accompanying identifying numbering system. Whenever possible, the names were taken from the front panel of the Happy Meal box, where the words, "Happy Meal" are displayed.

Numbering System - The numbering system reflects four aspects of the item:

Country of Origin/Country whose name is marked on the item. In some cases, the country of distribution is the country of origin; in most cases the country of origin is not the country of distribution. Many European Happy Meal premiums and Happy Meal boxes and bags are made in Germany and not specifically distributed in Germany and/or other parts of Europe. If the box, toy, or item is marked "Made in Germany," it is listed as GER for Germany with the accompanying date. Note that "***" denotes multiple country distribution.

Happy Meal Alphabetic code; two letters. In most cases, this code is made up of the first two letters of the Happy Meal name.

Year of distribution is the first year the item was distributed; dates vary among the countries listed.

Numerical listing of item; each general item is assigned a numerical number;

Example: **Can Ho7501 = Can**ada, **Ho**norary Meal of McDonaldland Promotion, 1975, Item **#01**

Can = Canada (country of origin)
Ho = Honorary Meal of McDonaldland Promotion
75 = 1975 (year of distribution)
01 = Numerical Listing of Item (i.e., first item given out)

Numerical designator - Last two numbers of identification code; whenever possible the following last two numbers have been used:

TOYS/PREMIUMS-	1-19
HM BOXES	20-25, 27-29
DISPLAY	26
COLOR CARD	27
LINE ART CARD-	28
HM BAG	30
CEILING DANGLER	41
COUNTER CARD	42
CREW CARD	43
CREW POSTER	44
REGISTER TOPPER	5
BUTTON	50
TRAYLINER	55
TABLE TENT	56
COUNTER MAT	60
MESSAGE CENTER INSERT	61
HEADER CARD-	62
LUG-ON	63
TRANSLITE/SMALL	64
TRANSLITE/LARGE	65
PIN	95

Whenever conflict in selecting the alpha/numeric designator arose, the first letter of the first two names of the Happy Meal was used and/or the generic alphabetic representation/combination of letters representing the item was used. For example, Batman, The Animated Series became Bt. Similarly, some Happy Meal promotions seem to be repeated over the years. These are consistently

assigned alphabetic listings, for example Attack Pack = AP; Barbie = BA; Cabbage Patch = CP; Funny Fry Friends = FF; Halloween = HA; Hot Wheels = HW, and Tonka = TK. As time progresses, it is hoped these alpha/numeric listings will become standard. The authors apologize for all past inconsistencies in developing a system which identifies each and every item from each and every country with an individual alpha/numeric label.

McDonald's Collecting Language

HM = Happy Meal
MIP = Mint in Package
MIT = Mint on Tree/Mint o Plastic Holder
MOC = Mint on Card
ND = No date listed on item
NP = Not packaged/given out in loose form
JT = Junior Tute/Happy Meal/German

Clean-up Week - open time period following a Happy Meal when no specific designated toy is distributed. The stock room backlog is given out in no particular order.

Counter Card - advertising or customer information card or board which sits on the counter.

Display - advertising medium which holds/displays the toys being promoted and distributed during a specific time frame. These range from older bubble type displays to 1990s cardboard fold-up type. They are displayed in stand-up Ronald McDonald cases and/or in the lobby display holder.

Generic - item such as a box or a toy not specifically associated with a specific theme Happy Meal or promotion. The item(s) may be used in several different promotions over a period of time; it may appear and reappear.

Header card - used in older Happy Meal promotions as advertising on top of the permanent display or ceiling dangler to display Happy Meal boxes or toys.

Insert card - advertising card within/along with the premium packaging.

Lug-on - sign added to the menu board.

McDonaldland - imaginary place where Ronald McDonald's cast of characters live and play; a playland area.

National - all stores in the geographical area distribute the same Happy Meal at the same time; supported with national advertising.

Register Topper - advertising item placed on the top of the register; cardboard advertising sign.

Regional - geographic distribution limited to specific cities, states, or stores; restricted distribution area.

Self-liquidator - item intended to be sold over the counter which may or may not be included in the Happy Meal promotion.

Table tent - rectangular shaped advertising sign placed on the tables and counters in the lobby.

Translite - transparent advertising sign used on overhead or drive-thru menu boards to illustrate the current promotion.

U-3 or U-5 - toys or premiums specifically designed for children under the age of 3 or 5; packaging is typically in zebra stripes around the outside of the package. The colors of the zebra stripes vary; typically a soft rubber toy.

This text encompasses a sampling of the Happy Meal toys distributed around the world during the 1995-99 time frame. Considering the fact that McDonald's is located within 114 different countries and that some Happy Meal sets in the last four years have had over eighty toys per set, it is easy to realize the geometric progression of Happy Meal Toys Around the World. This text covers the best of the sets distributed during the last four years. It is our hope that collectors will collect those sets they enjoy and continue to have fun collecting McDonald's Happy Meal Toys Around the World, toys which are still relatively inexpensive. Happy Meal toys from the USA represent Americana, just as toys from the other 113 countries represent a time period in the history of that country. Remember, all of this collecting is designed to be in search of adventure and fun! The challenge is to see how many different sets you can put together in a reasonable fashion. The golden rule is, "HAVE FUN COLLECTING!"

1995

- **"Have You Had Your Break Today?"**

- **McDonald's 40th Year Anniversary (1955-1995)**

Airport / McDonaldland Happy Meal, 1995

Hol Ai9501 **Birdie in Helicopter** - Red/Yel Helicopter.
$3.00-4.00

Hol Ai9502 **Grimace Piloting Airplane** - Yel/Red Airplane.
$3.00-4.00

Hol Ai9503 **Hamburglar in Utility Truck** - Red Truck.
$3.00-4.00

Hol Ai9504 **Ronald in Baggage Loader** - Red/Yel Baggage
Tram Loader W Blue Ladder. $3.00-4.00

Jpn Ai9501 **Ronald in Car** - Red Car. $3.00-4.00

Comments: National Distribution: Holland - March/April 1995; UK
- 1995; Japan - 1995.

SoA Ai9546. South
American Blue Book.

Hol Ai9501

Hol Ai9502

Ger Ai9564. German Translite.

Hol Ai9503

Hol Ai9504

Generic Merchandiser
Display Unit Blue Book.

Fra Ai9526. French Display.

Amazing Wildlife/Zoo Plush/Na Zoowas/MacZoo Happy Meal, 1995/1997

❑ ❑ Can Am9501 **Asiatic Lion** - Beige/Tan Stuffed Lion.
$2.00-3.00

❑ ❑ Can Am9502 **Chimpanzee** - Brn/Tan Stuffed Chimpanzee.
$2.00-3.00

❑ ❑ Can Am9503 **African Elephant** - Gry Stuffed Elephant.
$2.00-3.00

❑ ❑ Can Am9504 **Koala** - Tan/Wht Stuffed Bear. $2.00-3.00
❑ ❑ Can Am9505 **Dromedary Camel** - Brn Stuffed Camel.
$2.00-3.00

❑ ❑ Can Am9506 **Galapagos Tortoise** - Grn Stuffed Tortoise.
$2.00-3.00

❑ ❑ Can Am9507 **Polar Bear** - Wht Stuffed Bear. $2.00-3.00
❑ ❑ Can Am9508 **Siberian Tiger** - Gold/Blk/Wht Stuffed Tiger.
$2.00-3.00

❑ ❑ USA Am9701 **#1 Panda** - Blk/White Stuffed Bear.
$2.00-3.00

❑ ❑ USA Am9702 **#2 Rhinoceros** - Grey Rhino W White Horn/
Stuffed. $2.00-3.00

❑ ❑ USA Am9703 **#3 Yak** - Brn Yak W Wht Horns/Stuffed.
$2.00-3.00

❑ ❑ USA Am9704 **#4 Moose** - Tan Moose W Wht Antlers/Stuffed.
$2.00-3.00

❑ ❑ USA Am9705 **#5 Gorilla** - Black/Stuffed. $2.00-3.00
❑ ❑ USA Am9706 **#6 Bear** - Tan/Stuffed. $2.00-3.00

Comments: National Distribution: Canada - April 1-28, 1995. The advertising tie-in partner was the Canadian Wildlife Federation. Promotion included promo ad for Ranger Rick Magazine. Distribution: New Zealand - 1996 when the set of eight was distributed in two series of four at two different times, called Amazing Wildlife the first time and Zoo Plush the second time. The second set of four were called "Animal Pals" and distributed in the USA in 1997. Around the world different selections were distributed, four of the eight in most cases. Germany (#1, 5, 7, 8) in November 1996.

USA Am9701-06

CAN Am9501-04   CAN Am9505-08

Jpn Am9526. Japanese Display.

German Insert Card.

Animaniacs II/Animaniacs Popoid Stretchers Happy Meal, 1995

- ❏ ❏ USA An9501 **Pinky & the Brain** - in Pnk Vehicle Lifting Blue Ball. $2.00-3.00
- ❏ ❏ USA An9502 **Goodfeathers** - in Red Truck W Yellow Key. $2.00-3.00
- ❏ ❏ USA An9503 **Dot & Ralph** - in Blue/Red Vehicle. $2.00-3.00
- ❏ ❏ USA An9504 **Wakko & Yakko** - in Blue Launcher Vehicle. $2.00-3.00
- ❏ ❏ USA An9505 **Slappy & Skippy** - in Yellow Helicopter. $2.00-3.00
- ❏ ❏ USA An9506 **Mindy & Buttons** - in Blue Baby Carriage. $2.00-3.00
- ❏ ❏ USA An9507 **Wakko, Yakko & Dot** - in Red/Yellow Rocket. $2.00-3.00
- ❏ ❏ USA An9508 **Hip Hippos** - in Red Boat. $2.00-3.00

Comments: Distribution: USA - November 1-30, 1995; Belgium - 1996. Premium Markings "1995 Warner Bros;" Latin America (#1, 2, 4, 7) in April 1996.

USA An9501-04

USA An9505-08

Barbie/Hot Wheels Happy Meal, 1995

- ❏ ❏ Jpn Ba9306 **Secret Heart Barbie** - Wht/Rose Long Gown/ Holding Red Heart/Long Blond Hair. $2.00-4.00

- ❏ ❏ Uk Ba9302 **Hollywood Hair Barbie** - Gold Short Dress/Blu Star Base. $2.00-4.00
- ❏ ❏ Uk Ba9307 **Twinkle Lights Barbie** - Pink/Wht Gown/Wht Purse/Long Blonde Syn Hair. $2.00-4.00

- ❏ ❏ Jpn Ba9403 **Camp Barbie** - Pnk Jacket/Blonde Hair/Blu Shorts/Grn Base. $2.00-4.00
- ❏ ❏ Jpn Ba9407 **Jewel/Glitter Bride** - Wht Long Dress/Blonde Hair/Pnk Flowers. $2.00-4.00

- ❏ ❏ Jpn Hw9309 **McD Funny Car** - Red/Wht/"McDonald's" on Side . $2.00-4.00
- ❏ ❏ Jpn Hw9312 **Hot Wheels Funny Car** - Wht/Red/Yel "Hot Wheels" on Side Funny Car. $2.00-4.00

- ❏ ❏ Jpn Hw9413 **Turbine 4-2 Car** - Blu Turbine/Jet Car. $2.00-4.00

- ❏ ❏ Jpn Hw9517 **#27 Hot Wheels Car** - Blu W Yel McD Logo/ Yel #27 on Hood. $3.00-5.00

- ❏ ❏ Uk Hw9310 **Quaker State Racer #62 Car** - Grn Quaker State #62. $2.00-4.00
- ❏ ❏ Uk Hw9312 **Hot Wheels Funny Car** - Wht/Red/Yel "Hot Wheels" on Side Funny Car. $2.00-4.00

Comments: National Distribution: UK - August, 1995; Japan - 1995. UK Ba9307/10 and Hw9310/12 = USA Ba9307/10 and Hw9310/12 (See USA Barbie/Hot Wheels Happy Meal, 1993). The USA distributed eight Barbies and eight Hot Wheels during the 1993 Happy Meal Promotion. Selections from prior USA 1993 and 1994 promotions were combined around the world.

Barbie/Hot Wheels VI Happy Meal, 1995

- ❏ ❏ USA Ba9501 **#1 Hot Skatin' Barbie** - Turq/Pnk Outfit W Yel Skates. $2.00-3.00
- ❏ ❏ USA Ba9502 **#2 Dance Moves Barbie** - Pnk/Yel Tutu/Yel Shoes W Grn Stand/2p. $2.00-3.00

- ❏ ❏ USA Ba9503 **#3 Butterfly Princess Teresa** - Pnk Long Dress **W SOLID COLORS CUTOUTS**. $2.00-3.00
- ❏ ❏ USA Ba9509 **#3 Butterfly Princess Teresa** - Pnk Long Dress **W WAVEY COLORS CUTOUTS**. $2.00-3.00

- ❏ ❏ USA Ba9504 **#4 Country Barbie** - Purp/Pnk Cowgirl Barbie Riding Beige Horse. $2.00-3.00

- ❏ ❏ USA Ba9505 **#5 Lifeguard Ken - White Ken** W Yel Jet Ski/ 2p. $2.00-3.00
- ❏ ❏ USA Ba9519 **#5 Lifeguard Ken - Black Ken** W Yel Jet Ski/ 2p. $4.00-5.00

- ❏ ❏ USA Ba9506 **#6 Lifeguard Barbie - White Barbie**/Red-Wht-Blu Outfit Holding Blk Binoculars. $2.00-3.00
- ❏ ❏ USA Ba9520 **#6 Lifeguard Barbie - Black Barbie**/Red-Wht-Blu Outfit Holding Blk Binoculars. $4.00-5.00

- ❏ ❏ USA Ba9507 **#7 Bubble Angel Barbie** - Lt/Dk Blu Wrap Dress/Lt Blu Butterfly Wings/Bubble Holes. $2.00-3.00

USA Ba9501-04

❑ ❑ USA Ba9508 **#8 Ice Skatin' Barbie** - Blk Barbie/Turq/Pnk Outfit/Pnk/Sil Skates/Lt Blue Stand/2p. $2.00-3.00

❑ ❑ Lam Ba9617 **#17 Whale Watching Barbie** - Barbie W Binoculars on Stand W Whale. $2.00-4.00

❑ ❑ USA Hw9509 **#9 Lightning Speed** - Org/Blu W Clear Dome Cover. $2.00-2.50

❑ ❑ USA Hw9510 **#10 Shock Force** - Black Hot Rod W Yel Center/Top W Sil Pipes. $2.00-2.50

❑ ❑ USA Hw9511 **#11 Twin Engine** - Grn W Purp/Silver Engines/Blu Accents. $2.00-2.50

❑ ❑ USA Hw9512 **#12 Radar Racer** - Blu/Purp W Clear Dome. $2.00-2.50

❑ ❑ USA Hw9513 **#13 Blue Bandit** - Blu W Blk/Silver Accents. $2.00-2.50

❑ ❑ USA Hw9514 **#14 Power Circuit** - Red/Yel W Clear Dome. $2.00-2.50

❑ ❑ USA Hw9515 **#15 Black Burner** - Burg Red W Sil/Blk Accents. $2.00-2.50

❑ ❑ USA Hw9516 **#16 After Blast** - Pea Grn W Clear Dome Windshield/Blk Accents. $2.00-2.50

Comments: Distribution: USA - August 1-28, 1995; Latin America (#1, 2, 3, 17) and (# USA Hw9606, 07, 09, 10); Japan - 1996. Selections from two different USA series were combined in the Latin American market (Lam). All the Barbies have "real hair."

USA Ba9505-08

USA Ba9506, 08, 09, 20

USA Hw9509-12

USA Hw9513-16

Two U-3 premiums and USA Ba 9504

Four Barbies and four Hot Wheels. Japan, 1996.

Fra Ba9564

1995

Batman Badges Happy Meal, 1995

❑ ❑   Aus Bt9501 **Badge - Two Face, xographic.** $4.00-5.00
❑ ❑   Aus Bt9502 **Badge - Robin, xographic.** $4.00-5.00
❑ ❑   Aus Bt9503 **Badge - Batman, xographic.** $4.00-5.00
❑ ❑   Aus Bt9504 **Badge - Riddler, xographic.** $4.00-5.00

Comments: Distribution: Australia - July, 1995.

Aus Bt9501-04

Batman Cards Happy Meal, 1995

❑ ❑   USA Bt9501 **Batman Card Packet #1 - four cards.**
     $4.00-5.00

❑ ❑   USA Bt9502 **Batman Card Packet #2 - four cards.**
     $4.00-5.00

❑ ❑   USA Bt9503 **Batman Card Packet #3 - four cards.**
     $4.00-5.00

❑ ❑   USA Bt9504 **Batman Card Packet #4 - four cards.**
     $4.00-5.00

Comments: Distribution: New Zealand - 1995.

USA Bt9501-04

Bead Game Promotion, 1995

❑ ❑   Zea Be9501 **Bead Game Birdie** - Pnk/Circle.
     $3.00-4.00

❑ ❑   Zea Be9502 **Bead Game Grimace** - Yel/Circle.
     $3.00-4.00

❑ ❑   Zea Be9503 **Bead Game Hamburglar** - Grn/Circle.
     $3.00-4.00

❑ ❑   Zea Be9504 **Bead Game Ronald** - Red/Circle.
     $3.00-4.00

Comments: Regional Distribution: New Zealand - 1995.

Zea Be9501-03

Bubble Blowers Happy Meal, 1995

❑ ❑   Aus Bu9501 **Bubble Blower -** Birdie in red container.
     $3.00-4.00

❑ ❑   Aus Bu9502 **Bubble Blower -** Ronald in yellow container.
     $3.00-4.00

❑ ❑   Aus Bu9503 **Bubble Blower -** Grimace in purple container.
     $3.00-4.00

❑ ❑   Aus Bu9504 **Bubble Blower -** Hamburglar in green container.
     $3.00-4.00

Comments: Distribution: Australia - June/July/August, 1995; Middle East (except Israel); North Africa - June/July, 1995.

Busy World of Richard Scarry Happy Meal, 1995

❑ ❑   USA Ri9501 **Lowly Worm Red Apple Vehicle W Blue Post Office -** Red Apple/Blue Post Office/Cardboard/3p. $2.00 -3.00

❑ ❑   USA Ri9502 **Huckle Cat in Blue Vehicle W Yellow School -** Cat in Blu Car/Yel School/Cardboard Photo/3p. $2.00-3.00

❑ ❑   USA Ri9503 **Mr. Frumble in Green Vehicle W Red Fire Station -** in Grn Car/Red Fire Station/Cardboard/3p. $2.00-3.00

❑ ❑   USA Ri9504 **Banana Gorilla in Yellow Vehicle W Green Grocery Store -** in Yel Car/Grn Grocery Store/C Board/3p. $2.00-3.00

Comments: National Distribution: USA - September 1 - 30, 1995.

USA Ri9501-04

Cabbage Patch Kids Happy Meal, 1995

❏ ❏   Can Cp9301 **CP Jennifer Rita Tiny Dancer** - Purp/Wht Ballet Dress. $2.00-3.00

❏ ❏   Can Cp9302 **CP Christina Maria Happy Birthday** - Pnk Dress/Grn Gift/**No Teddy Bear**. $2.00-3.00

❏ ❏   Can Cp9303 **CP Jaime Christine Fun on Ice** - Wht Doll/ Wht Muff/**No Holly**/Turq Dress/Wht Skates. $2.00-3.00

❏ ❏   Can Cp9304 **CP Emily Elizabeth Sweet Dreamer** - Pnk Night Gown/Teddy Bear/**No Stocking**. $2.00-3.00

❏ ❏   USA Cp9401 **Mimi Kristina All Dressed Up Angel** - with Gold Horn. $2.00-3.00

❏ ❏   USA Cp9402 **Kimberly Katherine Santa's Helper** - with white Apron. $2.00-3.00

❏ ❏   USA Cp9403 **Abigail Lynn Toy Soldier -** wearing a Blue Top Hat/Candy Cane/Blk Doll. $2.00-3.00

❏ ❏   USA Cp9404 **Michelle Elyse Snow Fairy -** wearing a White Dress/Snowflake/Wht Doll. $2.00-3.00

Comments: Distribution: New Zealand - 1995. *** Identical Toys: Can/Phi/Sin/Zea Cp9301/02/03/04. *** Identical Cabbage Patch Dolls: New Zealand, Canada, USA.

Can Cp9301-04

USA Cp9401-04

Character Mazes Pinball Games Happy Meal, 1995

❏ ❏   Zea Ma9501 **Pinball Maze:** Birdie with Pink Background. $3.00-5.00

❏ ❏   Zea Ma9502 **Pinball Maze:** Ronald with Green Background. $3.00-5.00

❏ ❏   Zea Ma9503 **Pinball Maze:** Hamburglar with Purple Background. $3.00-5.00

❏ ❏   Zea Ma9504 **Pinball Maze:** Grimace with Purple Background. $3.00-5.00

Comments: Distribution: New Zealand - 1995. Mazes are miniature and rectangular shaped with pinball features in the plastic cases.

Zea Ma9501-04

Character Puzzle Happy Meal, 1995

❏ ❏   Zea Pu9501 **Puzzle: Birdie with Violin.** $3.00-5.00
❏ ❏   Zea Pu9502 **Puzzle: Ronald Playing Drums.** $3.00-5.00
❏ ❏   Zea Pu9503 **Puzzle: Hamburglar on Sax.** $3.00-5.00
❏ ❏   Zea Pu9504 **Puzzle: Grimace Singing.** $3.00-5.00

Comments: Distribution: New Zealand - 1995.

Character Straws Happy Meal, 1995

❏ ❏   Zea St9501 **Birdie Straw - pink or yellow.** $3.00-5.00
❏ ❏   Zea St9502 **Ronald Straw - red or green.** $3.00-5.00
❏ ❏   Zea St9503 **Grimace Straw - red or green.** $3.00-5.00
❏ ❏   Zea St9504 **Hamburglar Straw - pink or yellow.** $3.00-5.00

Comments: New Zealand 1995.

Zea St9501-03

Darkwing Duck I, 1995

❑ ❑ Aus Da9301 **Darkwing Duck** - in 3 Wheel Duckplane.
$3.00-4.00
❑ ❑ Aus Da9302 **Honker Muddle Foot** - in Can. $3.00-4.00
❑ ❑ Aus Da9303 **Launchpad McQuack** - in 2-Wheel Duckplane.
$3.00-4.00
❑ ❑ Aus Da9304 **Gosalyn** - in 2-Wheel Car. $3.00-4.00

Comments: Distribution: New Zealand - 1995. *** Identical Toys:
Aus/Zea Da9301/02/03/04

Aus Da9301-04

Disney Fun Rides Happy Meal, 1995

❑ ❑ Zea Di9301 **Donald in a Pirate's Ship Boat** - in Wht Boat/
Ship. $4.00-5.00
❑ ❑ Zea Di9302 **Goofy in Train Engine** - on Top of Red Engine.
$4.00-5.00
❑ ❑ Zea Di9303 **Mickey in Fire Engine** - in Red Fire Engine.
$4.00-5.00
❑ ❑ Zea Di9304 **Minnie in Tea Cup and Saucer** - in Yel Tea
Cup/Spins. $4.00-5.00

Comments: National Distribution: New Zealand - December 11,
1993 and 1995; Japan - August 1993; Singapore - October-November,
1994. Vehicles are similar to Euro Disney vehicles.

Zea Di9301-04

Disney Video Calendar Happy Meal, 1995

❑ ❑ Jpn Vi9501 **1996 Calendar** - 101 Dalmatians. $5.00-8.00
❑ ❑ Jpn Vi9502 **101 Dalmatians in a Box** - Happy Birthday Train
car. $2.00-4.00
❑ ❑ Jpn Vi9503 **Simba** - Lion King Figure. $2.00-3.00
❑ ❑ Jpn Vi9504 **Pongo** - 101 Dalmatian dog from 1991 USA
promo. $2.00-4.00

Comments: Distribution: Japan - November/December 1995.

Disneyland Adventures 40 Years Happy Meal, 1995

❑ ❑ Can Di9501 **Brer Bear on Splash Mountain** - in Brn Log
Boat/Viewer. $2.00-3.00
❑ ❑ Can Di9502 **Aladdin & Jasmine at Aladdin's Oasis** - on
Elephant/Purp Viewer. $2.00-3.00
❑ ❑ Can Di9503 **Simba in the Lion King Celebration** - on Rock/
Mountain/Brn Viewer. $2.00-3.00
❑ ❑ Can Di9504 **Mickey Mouse on Space Mountain** - in Red
Space Car/Viewer. $2.00-3.00

❑ ❑ Can Di9505 **Roger Rabbit in Mickey's Toontown** - in Yel/
Blu Car/Viewer. $2.00-3.00
❑ ❑ Can Di9506 **Winnie/Pooh on Big Thunder Mountain** -
Red Train W Blk or Grn Cab/Viewer. $2.00-3.00
❑ ❑ Can Di9507 **Peter Pan in Fantasmic!** - in Org Boat/Viewer.
$2.00-3.00
❑ ❑ Can Di9508 **King Louie on the Jungle Cruise** - in Grn/Yel
Jungle Boat/Viewer. $2.00-3.00

❑ ❑ Can Di9509 **U-3 Winnie the Pooh/Thunder Mountain** -
in Train W Grn Cab/No Viewer. $2.00-3.00

Comments: National Distribution: Canada - June 1-31, 1995. Can
Di9509 and Can Di9506 are not the same; Can Di9506 has Viewer (Cab
in green and/or black version with viewer) while U-3 green version does
not have viewer (3 toys in all).

Can Di9501-04

Can Di9505-08

Dolls Happy Meal, 1995

❏ ❏ Jpn Do9501 **Doll - Birdie** - 2 1/2" W Key Ring.
$3.00-4.00

❏ ❏ Jpn Do9502 **Doll - Grimace** - 2 1/2" W Key Ring.
$3.00-4.00

❏ ❏ Jpn Do9503 **Doll - Hamburglar** - 2" W Key Ring.
$3.00-4.00

❏ ❏ Jpn Do9504 **Doll - Ronald** - 3" W Key Ring. $3.00-4.00

Comments: Regional Distribution: Japan - 1995 as self-liquidator.

Jpn Do9501, 9502, 9504

Drive and Fly/Speedies/Classics/McOmoshiro Drive Happy Meal, 1995

❏ ❏ *** Dr9501 **Airplane Fry Kid** - in Blu Plane. $3.00-4.00

❏ ❏ *** Dr9502 **Bus Hamburglar** - in Grn Double Decker Bus.
$3.00-4.00

❏ ❏ *** Dr9503 **Car Birdie** - in Silver Auto. $3.00-4.00

❏ ❏ *** Dr9504 **Car Ronald** - in Red Car W Logo on Hood.
$3.00-4.00

Comments: National Distribution: Holland - January 4-February 7, 1995; UK - December 15-January 15, 1995; Europe - March 1995. Called "Speedies" in Holland and "Drive and Fly Happy Meal" in the UK; "Classics" in other parts of Europe plus Mc Omoshiro Drive in Japan. *** Identical Toys: Hol/UK Dr9501/02/03/04.

Uk Dr9501-04

Fuzzies Promotion, 1995

❏ ❏ Jpn Fu9501 **Fuzzie Fry Girl**. $3.00-4.00
❏ ❏ Jpn Fu9502 **Fuzzie Grimace**. $3.00-4.00
❏ ❏ Jpn Fu9503 **Fuzzie Hamburglar**. $3.00-4.00
❏ ❏ Jpn Fu9504 **Fuzzie Birdie**. $3.00-4.00

Comments: Regional Distribution: Japan - 1995 as self-liquidator and/or sold in McDonald's retail shops in Japan.

Jpn Fu9501-04

Generic Promotions, 1995

❏ ❏ Can Ge9501 **Calendar '95** - Underwater World.
$2.00-3.00

❏ ❏ Net Ge9502 **Postcard Ronald** - Jumping thru Hoop.
$1.00-3.00

❏ ❏ Net Ge9503 **Game Find Way to McD** - Paper W Punchout Characters. $1.00-3.00

❏ ❏ Aus Ge9504 **Calendar '95** - Colossal Calendar W Grow Chart. $3.00-5.00

❏ ❏ Zea Ge9505 **#1 Activity Book/Coloring Book**.
$1.00-2.00

❏ ❏ Zea Ge9506 **#2 Book/Magic Dot-To-Dot**. $1.00-2.00
❏ ❏ Zea Ge9507 **#3 Trace & Color Book**. $1.00-2.00
❏ ❏ Zea Ge9508 **#4 Colour Book**. $1.00-2.00
❏ ❏ Zea Ge9509 **Calendar '95**-Amazing Animals. $3.00-5.00

Comments: Regional Distribution: Germany/Holland/Latvia - 1995; Australia, New Zealand - 1995.

Halloween '95/What Am I Going to Be for Halloween?/Ghosts and Monsters Happy Meal, 1995

❑ ❑ USA Ha9501 **Tape: Ronald Makes it Magic Cassette**.
$1.00-1.50

❑ ❑ USA Ha9502 **Tape: Travel Tunes Cassette**. $1.00-1.50

❑ ❑ USA Ha9503 **Tape: Silly Sing-Along Cassette**.
$1.00-1.50

❑ ❑ USA Ha9504 **Tape: Scary Sound Effects Cassette**.
$1.00-1.50

❑ ❑ USA Ha9505 **Hamburglar w Witch costume -** Blk Witch Snap-On/3p. $2.00-2.50

❑ ❑ USA Ha9506 **Grimace w Ghost costume -** Wht Ghost Snap-On/3p. $2.00-2.50

❑ ❑ USA Ha9507 **Ronald w Frankenstein costume -** Grn Frankenstein Snap-On/3p. $2.00-2.50

❑ ❑ USA Ha9508 **Birdie w Pumpkin costume -** Org Pumpkin Snap-On/3p. $2.00-2.50

❑ ❑ Hol Ha9509 **Birdie w Witch costume -** Org Pumpkin Snap-On Purple Witch Costume/3p. $4.00-6.00

Comments: National Distribution: USA - October 1-31, 1995. Called "Ghosts and Monsters" Happy Meal in Sweden. USA Ha9505-08 were distributed around the world. Additionally, Birdie as a witch premium was distributed in Holland.

Hol Ha9501-04

Purple Witch Birdie, Holland.

USA Ha9501-04

USA Ha9505-08

Hol Ha9509. Insert Card

24

Intergalactic Adventure Vehicles Happy Meal, 1995

☑ ❑ Uk In9501 **McRobot Man** - Blu/Yel Vehicle. $2.00-3.00
☑ ❑ Uk In9502 **Moon Buggy Grimace** - Purp W Grn Vehicle.
$2.00-3.00
☑ ❑ Uk In9503 **Space Shuttle Ronald/Grimace** - Wht Space-
ship W Blu Wehicle. $2.00-3.00
☑ ❑ Uk In9504 **Lunar Rover Ronald** - Yel Ron W Red Vehicle.
$2.00-3.00

Comments: National Distribution: UK - February/March, 1995;
Europe - May, 1995.

Uk In9501-04

Uk In9526

Lion King Stampers Happy Meal, 1995

❑ ❑ Aus Bt9501 **Stamper - Timon & Pumbaa**. $3.00-4.00
❑ ❑ Aus Bt9502 **Stamper - Simba & Nala**. $3.00-4.00
❑ ❑ Aus Bt9503 **Stamper - Scar**. $3.00-4.00
❑ ❑ Aus Bt9504 **Stamper - Mufasa**. $3.00-4.00

Comments: Australia - October, 1995.

Uk Intergalactic Vehicle Translite. Uk In9564.

Australian Lion King Counter Card.

1995

McDonaldland Sports Happy Meal, 1995

❑ ❑ Aus Sp9501 **Birdie with a Tennis Racket**. $4.00-5.00
❑ ❑ Aus Sp9502 **Ronald with a Soccer Ball**. $4.00-5.00
❑ ❑ Aus Sp9503 **Grimace playing Baseball**. $4.00-5.00
❑ ❑ Aus Sp9504 **Hamburglar Lifting Weights**. $4.00-5.00

Comments: Distribution: Australia - 1995.

Aus Sp9501-04

McFarm Happy Meal, 1995

🖼 ❑ Uk Fa9501 **Birdie/Wheelbarrow** - Pushing Brn Wheelbarrow W Ducks/2p. $3.00-5.00
🖼 ❑ Uk Fa9502 **Grimace/Pick-Up Truck** - in Red Pick-Up Truck. $3.00-5.00
🖼 ❑ Uk Fa9503 **Hamburglar/Harvester** - in Red Harvester. $3.00-5.00
🖼 ❑ Uk Fa9504 **Ronald/Tractor** - in Grn Tractor. $3.00-5.00

Comments: National Distribution: UK - 1995; Germany - June, 1995; Europe - October, 1995; South/Central America - October - December, 1995.

Uk Fa9501-04

Fra McFarm Translite

McJuegos Character Ball Mazes Happy Meal, 1995

❑ ❑ Zea Ba9501 **Ball Maze:** Birdie with Pink Background. $4.00-5.00
❑ ❑ Zea Ba9502 **Ball Maze:** Grimace with Green Background. $5.00-8.00
❑ ❑ Zea Ba9503 **Ball Maze:** Hamburglar with Orange Background. $4.00-5.00
❑ ❑ Zea Ba9504 **Ball Maze:** Ronald with Yellow Background. $4.00-5.00

Comments: Distribution: New Zealand - 1995. Mazes are miniature, hand held circular ball mazes. The green Grimace maze is the most difficult to obtain.

Zea Ba9501-04

McRodeo/Pony Express Happy Meal, 1995

❏ ❏   *** Ro9501 **Birdie with a Lasso** - W Beige Lasso.
$3.00-5.00

❏ ❏   *** Ro9502 **Grimace riding a Horse** - on Brn Horse.
$3.00-5.00

❏ ❏   *** Ro9503 **Hamburglar as a Cowboy** - Wearing Red Pants/ Blk Hat.
$3.00-5.00

❏ ❏   *** Ro9504 **Ronald in a Barrel** - in Yel Barrel.
$3.00-5.00

Comments: National Distribution: Germany, Panama, New Zealand, UK - May 5-June 1995; Japan - tested in November 1994; Germany - May, 1995; Mexico - March/April 1995; New Zealand and France - 1995. ***Identical Toys: Ger/Mex/Zea/UK Ro9501/02/03/04.

Ger Ro9501-04

Pony Express Display, France.

Ger Ro9526. German McRodeo Danglers.

Pony Express Translite, France.

McTurbo Happy Meal, 1995

❑	❑	Jpn Tu9501	**Hamburglar on Blue car.**	$4.00-5.00
❑	❑	Jpn Tu9502	**Birdie on pink car.**	$4.00-5.00
❑	❑	Jpn Tu9503	**Ronald on red car.**	$4.00-5.00
❑	❑	Jpn Tu9504	**Grimace on green car.**	$4.00-5.00

Comments: Distribution: Japan - October/November, 1995.

Jpn Tu9501-04

Japanese Poster for
McTurbo.

Mickey's Toontown Finger Puppets & Activity Box Happy Meal, 1995

❑	❑	Aus Pf9501	**Finger Puppet - Mickey Mouse.**	$3.00-4.00
❑	❑	Aus Pf9502	**Finger Puppet - Minnie Mouse.**	$3.00-4.00
❑	❑	Aus Pf9503	**Finger Puppet - Goofy.**	$3.00-4.00
❑	❑	Aus Pf9504	**Finger Puppet - Donald Duck.**	$3.00-4.00

Comments: Distribution: Australia - September, 1995. Each finger puppet came with a gray and white Activity Box.

Australian Display for
Mickey's Toontown.

Monster Play-Doh Happy Meal, 1995

❑	❑	Ger Mo9501	**Play-Doh Canister with Mold - Yellow.**
			$2.00-4.00
❑	❑	Ger Mo9502	**Play-Doh Canister with Mold - Green.**
			$2.00-4.00
❑	❑	Ger Mo9503	**Play-Doh Canister with Mold - Blue.**
			$2.00-4.00
❑	❑	Ger Mo9504	**Play-Doh Canister with Mold - Red.**
			$2.00-4.00

Comments: Distribution: Germany - 1995. Each canister came with a mold.

Ger Mo9502

Muppet Workshop Happy Meal, 1995

❑	❑	Can Mu9505	**U-3 What-Not** - Yel Monster/Purp Cowboy Hat/Red Guitar/4p.	$2.00-3.00
❑	❑	*** Mu9501	**Wk 1 Bird** - Turq Bird/Red Hat/Purp Bow/4p.	$2.00-3.00
❑	❑	*** Mu9502	**Wk 2 Dog** - Pnk Dog/Org Bird Hat/Grn Camera/4p.	$2.00-3.00
❑	❑	*** Mu9503	**Wk 3 Monster** - Grn Monster/Org Hat/Blu Bear/4p.	$2.00-3.00
❑	❑	*** Mu9504	**Wk 4 What-Not** - Yel Monster/Purp Cowboy Hat/Red Guitar/4p.	$2.00-3.00

Comments: National Distribution: Canada - March, 1995; Argentina, Costa Rica, Mexico, Panama, Uruguay, Venzuela - March, 1995. *** Identical Toys: Can/Mex Mu9501/02/03/04.

USA Mu9501-04

USA Mu9564.
Translite.

Generic Mystery
Riders Blue Book.

Musical Tapes Happy Meal, 1995

❏	❏	Zea Ta9501 **Tape: ABCs #1**.	$4.00-5.00
❏	❏	Zea Ta9502 **Tape: ABCs #2**.	$4.00-5.00
❏	❏	Zea Ta9503 **Tape: ABCs #3**.	$4.00-5.00
❏	❏	Zea Ta9504 **Tape: ABCs #4**.	$4.00-5.00

Comments: Distribution: New Zealand - 1995.

Mystery Riders Happy Meal, 1995

❏	❏	Zea My9501 **#1 Green Helicopter**.	$4.00-5.00
❏	❏	Zea My9502 **#2 Red Choo Choo Train**.	$4.00-5.00
❏	❏	Zea My9503 **#3 Blue Steamboat**.	$4.00-5.00
❏	❏	Zea My9504 **#4 Pink & Purp Car**.	$4.00-5.00

Comments: National Distribution: New Zealand - January 1993 and 1995. By spinning the rectangle, different character faces appear.

NBA Basketball Tumblers Happy Meal, 1995

❏	❏	Zea Tu9501 **Lg Tumbler: Chicago Bulls**.	$2.00-3.00
❏	❏	Zea Tu9502 **Lg Tumbler: Detroit Pistons**.	$2.00-3.00
❏	❏	Zea Tu9503 **Lg Tumbler: Houston Rockets**.	$2.00-3.00
❏	❏	Zea Tu9504 **Lg Tumbler: Phoenix Suns**.	$2.00-3.00
❏	❏	Zea Tu9505 **Med Tumbler: Charlotte Hornets**.	$2.00-3.00
❏	❏	Zea Tu9506 **Med Tumbler: Houston Rockets**.	$2.00-3.00
❏	❏	Zea Tu9507 **Med Tumbler: Orlando Magic**.	$2.00-3.00
❏	❏	Zea Tu9508 **Med Tumbler: Phoenix Suns**.	$2.00-3.00
❏	❏	Zea Tu9509 **Med Tumbler: New York Knicks**.	$2.00-3.00
❏	❏	Zea Tu9510 **Med Tumbler: Charlotte Hornets**.	$2.00-3.00

Comments: Distribution: New Zealand - 1995.

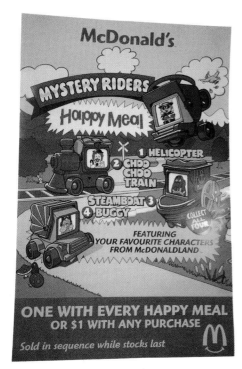

Zea My9501-04

Zea Tu9501-04

101 Dalmations Happy Meal, 1995

❑ ❑ Uk On9101 **Set 1 Pongo the Dog** - Blk/Wht Dalmation Standing. $2.00-3.00

❑ ❑ Uk On9502 **Set 2 Lucky the Pup** - Blk/Wht Dalmation Pup Sitting W Red Collar. $2.00-3.00

❑ ❑ Uk On9103 **Set 3 Colonel/Sgt. Tibs** - Sheep Dog W Cat. $2.00-3.00

❑ ❑ Uk On9104 **Set 4 Cruela De Vil** - Yel/Blk Villainess. $2.00-3.00

❑ ❑ Aus On9505 **Patch/Blanket** - Dalmation on Yel Blanket. $4.00-5.00

Comments: Regional Distribution: Australia - April, 1995; National Distribution: Hol/UK - January, 1995. Sets 1/2/3/4 = USA On9100-03.

Pan Pipes Happy Meal, 1995

❑ ❑ Zea Pi9501 **Pan Pipe Section - Blue, Green & Yellow**. $3.00-5.00

❑ ❑ Zea Pi9502 **Pan Pipe Section - Aqua, Yellow & Red**. $3.00-5.00

❑ ❑ Zea Pi9503 **Pan Pipe Section - Purple & Pink**. $3.00-5.00

Comments: Distribution: New Zealand. The musical pan pipe is made up of three sections, which fit together to make a Pan's Pipe. The sections were not widely distributed, making them a rare premium.

Placemats/Safety Happy Meal, 1995

❑ ❑ Zea Pl9501 **Placemat - Name the Sea Life**. $2.00-3.00
❑ ❑ Zea Pl9502 **Placemat - Someone....** $2.00-3.00
❑ ❑ Zea Pl9503 **Placemat - Keep Alert....** $2.00-3.00

Comments: Distribution: New Zealand - 1995 in a few markets. Placemats were distributed around the world in various markets when the planned Happy Meal premiums were in short supply.

Zea Pl9501-03

Pocahontas Finger Puppets Happy Meal, 1995

❑ ❑ Aus Pf9501 **Finger Puppet - Pocahontas**. $3.00-4.00
❑ ❑ Aus Pf9502 **Finger Puppet - Captain John Smith**. $3.00-4.00

❑ ❑ Aus Pf9503 **Finger Puppet - John Radcliff**. $3.00-4.00
❑ ❑ Aus Pf9504 **Finger Puppet - Meeko**. $3.00-4.00

Comments: Distribution: Australia - September 1995.

Aus Pf9501-04

Pocahontas II Happy Meal, 1995

❑ ❑ Uk Pc9501 **Pocahontas in Canoe**. $4.00-5.00
❑ ❑ Uk Pc9502 **Capt John Smith with Clip On Clothes**. $4.00-5.00

❑ ❑ Uk Pc9503 **John Radcliff & Percy**. $4.00-5.00
❑ ❑ Uk Pc9504 **Grandma's Willow Tree**. $4.00-5.00
❑ ❑ UK Pc9505 **Meeko Insider**. $4.00-5.00
❑ ❑ Bel Pc9506 **Puzzle - Pocahontas by the River**. $4.00-7.00

Comments: National Distribution: UK and Europe - 1995; Belgium - December, 1995.

Uk Pc9501-05

Hol Pc9501-05

Ger Po9565 German Translite.

French Translite.

Polly Pocket/Attack Pack/Excite Rider Happy Meal, 1996/1995

❑ ❑ Can Po9501 **Ring** - Pnk/Yel/Grn Polly Pocket on Flower Pedal.
$2.00-4.00

❑ ❑ Can Po9502 **Locket** - Pnk Heart Locket W Pnk Cord.
$2.00-4.00

❑ ❑ Can Po9503 **Watch** - Yel Case/Turq Dial. $2.00-4.00

❑ ❑ Can Po9504 **Bracelet** - Pnk/Turq Butterfly Case/Yel Strap.
$2.00-4.00

❑ ❑ Can Hw9505 **Truck** - Gry/Blk/Red Truck. $2.00-4.00

❑ ❑ Can Hw9506 **Battle Bird** - Grn/Wht Airplane/Bird.
$2.00-4.00

❑ ❑ Can Hw9507 **Lunar Invader** - Yel/Gry Lunar Module.
$2.00-4.00

❑ ❑ Can Hw9508 **Sea Creature** - Turq/Wht Sea Creature.
$2.00-4.00

Comments: National Distribution: Canada - February 3-March 2, 1995. Can Po9501-10 = USA Po9501-10; Japan - October/November, 1996.

USA Po9501-04

USA Hw9505-08

Power Rangers the Movie/Mighty Morphin Happy Meal, 1995

❏ ❏ USA Pr9501 **Power Com** - Grey Watch/Blk Strap/Flips Open.
$2.00-3.00
❏ ❏ USA Pr9502 **Powermorpher Buckle -** Grey/Red Buckle W
3 Gold Coins/4p. $2.00-3.00
❏ ❏ USA Pr9503 **Power Siren** - Blk/Grey Whistle. $2.00-3.00
❏ ❏ USA Pr9504 **Alien Detector** - Purple Case/Blue Door.
$2.00-3.00

Comments: National Distribution: USA - July 1-31, 1995.

Can Po9501-04

USA Pr9501-04

Puzzles Happy Meal, 1995

❏ ❏ Aus Pu9501 **Puzzle Birdie** - Skipping. $3.00-4.00
❏ ❏ Aus Pu9502 **Puzzle Grimace** - Exercising. $3.00-4.00
❏ ❏ Aus Pu9503 **Puzzle Hamburglar** - Scooting. $3.00-4.00
❏ ❏ Aus Pu9504 **Puzzle Ronald** - Skiing. $3.00-4.00

❏ ❏ Zea Pu9505 **Puzzle Grimace** - Dancing. $3.00-4.00
❏ ❏ Zea Pu9506 **Puzzle Hamburglar** - Playing Horn.
$3.00-4.00
❏ ❏ Zea Pu9507 **Puzzle Ronald** - Playing Drums. $3.00-4.00
❏ ❏ Zea Pu9508 **Puzzle Birdie** - Playing Violin. $3.00-4.00

Comments: Regional Distribution: Australia, New Zealand - 1995.
Numerous countries around the world gave out puzzles in place of
planned Happy Meal toys when supply ran out.

Jpn Po9526. Japanese Polly Pocket Display.

Aus Pu9501-04

Aus Pu9502

Zea Pu9505-08

Rip Racers Happy Meal, 1995

❏ ❏   Aus Ri9501 **Birdie in Yellow Racing Car - With Zip Strip**.
$3.00-5.00

❏ ❏   Aus Ri9502 **Grimace in Purple Racing Car - With Zip Strip**.
$3.00-5.00

❏ ❏   Aus Ri9503 **Hamburglar in Green Racing Car - With Zip Strip**.
$3.00-5.00

❏ ❏   Aus Ri9504 **Birdie in Red Racing Car - With Zip Strip**.
$3.00-5.00

Comments: Distribution: Australia - November 1995. Zip strip launches the plastic racing cars which have the characters illustrated on enclosed sticker sheets.

Sand Molds Happy Meal, 1995

❏ ❏   Zea Mo9501 **Sand Mold: Yellow House**. $2.00-3.00
❏ ❏   Zea Mo9502 **Sand Mold: Red Car**. $2.00-3.00
❏ ❏   Zea Mo9503 **Sand Mold: Green Train**. $2.00-3.00
❏ ❏   Zea Mo9504 **Sand Mold: Yellow Jumbo Jet**. $2.00-3.00

Comments: Distribution: New Zealand - 1995.

Snoopy and His Friends Happy Meal, 1995

❏ ❏   Jpn Sn9501 **Snoopy with Goggles - Goggles Change Color**.
$7.00-10.00

❏ ❏   Jpn Sn9502 **Snoopy with Typewriter**. $7.00-10.00
❏ ❏   Jpn Sn9503 **Snoopy with Flying Ace Hat**. $7.00-10.00
❏ ❏   Jpn Sn9504 **Snoopy with Boy Scout Hat & Backpack**.
$7.00-10.00

Comments: Distribution: Japan - 1995.

Jpn Sn9501-04

Sonic the Hedgehog World Happy Meal, 1995

❏ ❏ Jpn So9501 **Sonic the Hedgehog** - Blue & Orange, 2 pieces.
$2.00-4.00

❏ ❏ Jpn So9502 **Knuckles -** Red Figure On White Cloud.
$2.00-4.00

❏ ❏ Jpn So9503 **Dr. Egg Man/Dr. Ivo Robotnik** - Gray Auto With Figure. $2.00-4.00

Comments: National distribution: USA - 1994; Japan June/July, 1995; France - 1995.

France

Sonic Translite, France.

Ger So9526. German Display Danglers.

Space Rescue Happy Meal, 1995

❏ ❏ Can Sp9505 **U-3 Astro Viewer** - Grn/Purp W Wht Label.
$3.00-4.00

❏ ❏ Can Sp9501 **Astro Viewer** - Grn/Purp W Pnk Label.
$2.00-4.00

❏ ❏ Can Sp9502 **Tele Communicator** - Org/Grn. $2.00-4.00

❏ ❏ Can Sp9503 **Space Slate** - Blu/Purp/Org W Purp Pen/2p.
$2.00-4.00

❏ ❏ Can Sp9504 **Lunar Grabber** - Blu/Grn/Org. $2.00-4.00

Comments: National Distribution: Canada - March 3-31, 1995. Can Sp9501 = Can Sp9505, loose out of package. Can Sp9501-05 = USA Sp9501-05.

Speedsters/McDonaldland Launch Vehicles Happy Meal, 1995

❏ ❏ Uk Ss9501 **Birdie in Pink Jeep** - Purple Launcher w Garage/2 pieces. $4.00-5.00

❏ ❏ Uk Ss9502 **Grimace in Green Go-Kart** - Blue Launcher w Garage/2 pieces. $4.00-5.00

❏ ❏ Uk Ss9503 **Hamburglar on Red/Blue Bike** - Yellow Launcher w Garage/2 pieces. $4.00-5.00

❏ ❏ Uk Ss9504 **Ronald in Blue Boat** - Red Launcher w Boat House/2 pieces. $4.00-5.00

Comments: National Distribution: UK - 1995; Europe - 1996.

Spider-Man Happy Meal, 1995

❏ ❏   Can Sm9509 **U-3 Amazing Spiderman** - Red/Blu Spider-Man. $3.00-4.00

❏ ❏   Can Sm9501 **Amazing Spider-Man** - Red/Blu Spider-Man. $3.00-5.00

❏ ❏   Can Sm9502 **Scorpion Stingstriker** - Grn Scorpion Vehicle W Plier Claws. $3.00-5.00

❏ ❏   Can Sm9503 **Dr. Octopus** - Yel/Grn Man W Gry Tentacles. $3.00-5.00

❏ ❏   Can Sm9504 **Spider-Man Webrunner** - Spider-Man in Wht/Red/Blu Spider Vehicle. $3.00-5.00

❏ ❏   Can Sm9505 **Mary Jane Watson** - Pnk Coat/Yel Shirt/W Red or Grn Clip-On Dress/3p. $3.00-5.00

❏ ❏   Can Sm9506 **Venom Transport** - Blk/Wht/Red Spider Vehicle. $3.00-5.00

❏ ❏   Can Sm9507 **Spider-Sense Peter Parker** - Brn Shirt/Blu Pants/Half Face. $3.00-5.00

❏ ❏   Can Sm9508 **Hobgoblin Landglider** - Purp/Org/Gry Vehicle. $3.00-5.00

Comments: National Distribution: Canada: May: 1995. Can Sm9501 = Can Sm9509, loose out of package. Can Sm9501-09 = USA Sm9501-09. Spider-Man character names and likenesses are trademark and copyright: Marvel Entertainment Group, Inc.

USA Sm9501-04

USA Sm9505-08

Sports Phonecards Happy Meal, 1995

❏ ❏   Zea Ph9501 **Sport $5 Phone Card #1**. $4.00-7.00
❏ ❏   Zea Ph9502 **Sport $5 Phone Card #2**. $4.00-7.00
❏ ❏   Zea Ph9503 **Sport $5 Phone Card #3**. $4.00-7.00
❏ ❏   Zea Ph9504 **Sport $5 Phone Card #4**. $4.00-7.00

Comments: Distribution: New Zealand - 1995.

Summer Fun Happy Meal, 1995

❏ ❏   Aus Su9501 **Pail** - Red W Yel Handle. $2.00-3.00
❏ ❏   Aus Su9502 **Rake** - Purp. $2.00-3.00
❏ ❏   Aus Su9503 **Shovel** - Yel. $2.00-3.00
❏ ❏   Aus Su9504 **Mold-Star Fish** - Hot Pnk. $2.00-3.00

Comments: Regional Distribution: Australia - January (Summer), 1995.

Tiny Toon Adventures Happy Meal, 1995

❏ ❏   Can Ti9201 **Babs Bunny** - Pink Bunny W Tiny Toons Record Player in Bubble. $2.00-3.00

❏ ❏   Can Ti9202 **Buster Bunny** - Blu Bunny in Red Bumper Car/Basketball Bubble. $2.00-3.00

❏ ❏   Can Ti9203 **Dizzy Devil** - Pur Dizzy Devil in Bubble Car. $2.00-3.00

❏ ❏   Can Ti9204 **Elmyra** - Girl W Yel Hat in Grn Car W Bunny in Bubble. $2.00-3.00

Can Ti9201-04

❏ ❏   Can Ti9205 **Gogo Dodo** - Grn Gogo Dodo on Yel 3 Wheel Roller. $2.00-3.00

❏ ❏   Can Ti9206 **Montana Max** - Max in Grn Cash Register Car. $2.00-3.00

❏ ❏   Can Ti9207 **Plucky Duck** - Plucky in Blu Steam Roller Car. $2.00-3.00

❏ ❏   Can Ti9208 **Sweetie** - Pink Bunny on Pavement Roller. $2.00-3.00

Can Ti9205-08

❏ ❏ Zea Ti9509 **Booklet - Coloring Book.** $2.00-3.00
❏ ❏ Zea Ti9510 **Booklet - Dot-to-Dot.** $2.00-3.00
❏ ❏ Zea Ti9511 **Booklet - Activity Book.** $2.00-3.00
❏ ❏ Zea Ti9512 **Booklet - Trace N Color.** $2.00-3.00

Comments: National Distribution: Canada - October-November, 1992; Mexico, Costa Rica - January, 1992; Puerto Rico - February, 1991; New Zealand - 1995.

Zea Ti9509-12

USA Ti9264

Tiptoe Teddy Christmas Bears Happy Meal, 1986/1995

❏ ❏ Sin Be9501 **Brother Bear** - in Blue Jacket W Red Christmas Book. $7.00-10.00
❏ ❏ Sin Be9502 **Sister Bear** - in Pink Dress. $7.00-10.00
❏ ❏ Sin Be9503 **Father Bear** - W Green Tie. $7.00-10.00
❏ ❏ Sin Be9504 **Mother Bear** - W Red Dress. $7.00-10.00

Comments: Regional Distribution: Singapore distributed four bears called "Tip Toe Teddy." These were not the same bears as the 1986 Canadian promotion.

Sin Be9501-04

Totally Toy Holiday/Mattel Happy Meal, 1995

❏ ❏ USA To9501 **#1 Holiday Barbie -** Grn/Wht Dressed Barbie in Wht/Gold Sleigh. $2.50-4.00

❏ ❏ USA To9502 **#2 57 Chevy Hot Wheels - Red 57 Chevy Hw Car W Blue Ramp/2p.** $2.00-2.50
❏ ❏ USA To9512 **#2 Gator Hot Wheels - Grn Hw Car/Gator Graphics W Blue Ramp/2p.** $2.00-2.50

❏ ❏ USA To9503 **#3 Polly Pocket House -** Grn/Pnk/Wht House W Girl Fig Inside. $2.00-2.50
❏ ❏ USA To9504 **#4 Mighty Max Case -** Two Tone Blue Case/ MM Figurine Inside. $2.00-2.50
❏ ❏ USA To9505 **#5 Cabbage Patch Kids on Rocking Horse** - Pnk/Red Rocking Horse Case/Fig Inside. $2.00-2.50
❏ ❏ USA To9506 **#6 Hot Wheels: North Pole Explorer Vehicle -** Blu/Blk/Wht Explorer Truck/Purp Fig. $2.00-2.50
❏ ❏ USA To9507 **#7 Fisher-Price: Once upon a Dream Princess Fig -** Red/Wht Fig W Wht Crown. $2.00-2.50
❏ ❏ USA To9508 **#8 Fisher-Price: Great Adventure Knight Figurine W Green Dragon -** Blk/Red Shield Fig/2p. $2.00-2.50

Comments: National Distribution: USA - December 1-31, 1995.

Tricky Trackers/New Year Parade/Speed Macs/Carriles Locos/ McMobiles Happy Meal, 1995

❏ ❏ Sin Tr9501 **Track/Fries on Yellow Track.** $3.00-4.00
❏ ❏ Sin Tr9502 **Track/Big Mac on Orange Track.** $3.00-4.00
❏ ❏ Sin Tr9503 **Track/Hamburger on Red Track.** $3.00-4.00
❏ ❏ Sin Tr9504 **Track/Shake on Blue Track.** $3.00-4.00

Tricky Trackers Vehicles.

❏ ❏ Zea Tt9505 **Track/Hamburger on Orange Track - Hamburglar.** $3.00-4.00

❏ ❏ Zea Tt9506 **Track/Milkshake on Purple Track - Grimace.** $3.00-4.00

❏ ❏ Zea Tt9507 **Track/Fries on Yellow Track - Birdie.** $3.00-4.00

❏ ❏ Zea Tt9508 **Track/Cheeseburger on Red Track - Ronald.** $3.00-4.00

❏ ❏ Bra Tt9509 **Track/Fries on Turquoise Track - Birdie.** $3.00-4.00

❏ ❏ Bra Tt9510 **Hamburglar Burger on Yellow Track.** $3.00-4.00

Comments: Distribution: New Zealand - 1995; Latin America - February, 1996 (#5, 6); Germany - June, 1996; Singapore - February, 1995; Japan - tested in January and run in October, 1994; Argentina - May, 1994; Costa Rica, Panama, Venezuela, Chile - September, 1994. Each vehicle came with a character sticker plus scenery stickers for board. Tracks are interconnectible; connect together to create one larger track. The color of the track and the windup food toys changed from country to country.

Sin Tr9501, 03

Zea Tt9505-08

Hon Tr9501-04

Ze passen bij elkaar én in elkaar.

Het Tricky Trackers Happy Meal. 7,75.

Hol Tr9501-04

37

Ger Tr9526. German Display with Danglers.

Fra Tr9564. French translite.

Under the Big Top Happy Meal, 1994/1995

❑ ❑ Zea Un9401 **Ladder Birdie** - Birdie Acrobat/Flips/3p.
$3.00-5.00
❑ ❑ Zea Un9402 **Barrel Grimace** - Drum/Rolls. $3.00-5.00
❑ ❑ Zea Un9403 **Cannon Fry Kids** - Pump/Shoot/2p
$5.00-8.00
❑ ❑ Zea Un9404 **Elephant Ronald** - Balancing/Riding/2p.
$3.00-5.00

Comments: Regional Distribution: Hong Kong, Singapore, New Zealand - 1994; National Distribution: New Zealand - March, 1995. Zea Un9503 Fry Kids Cannon was recalled.

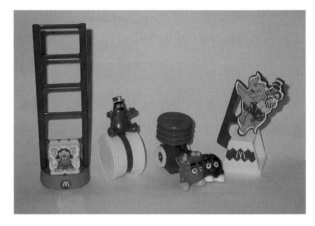

Zea Un9401-04

Vehicles Happy Meal, 1995

❑ ❑ Jpn Ve9501 **Birdie** - in Pink Windup Car. $3.00-5.00
❑ ❑ Jpn Ve9502 **Grimace** - in Red Windup Car. $3.00-5.00
❑ ❑ Jpn Ve9503 **Hamburglar** - in Yellow Windup Car.
$5.00-8.00
❑ ❑ Jpn Ve9504 **Ronald** - in Green Windup Truck. $3.00-5.00

Comments: Distribution: Japan - 1995.

Jpn Ve9526. Japanese Display.

Water Fun Happy Meal, 1995

❏	❏	Uk Wf9501 **Squirt Gun - Frog in Green**.	$2.00-4.00
❏	❏	Uk Wf9502 **Squirt Gun - Pig in Pink**.	$2.00-4.00
❏	❏	Uk Wf9503 **Squirt Gun - Shark in Blue**.	$2.00-4.00
❏	❏	Uk Wf9504 **Squirt Gun - Cat in Orange**.	$2.00-4.00

Comments: National Distribution: UK - July 1995.

Water Fun/Aqua Zoo Squirt Guns/Animal Water Pistals/Squirters Happy Meal, 1995

❏	❏	Bel Wa9501 **Squirt Gun - Toucan in Blue And Yellow**.	
			$2.00-3.00
❏	❏	Bel Wa9502 **Squirt Gun - Hippo in Pink**.	$2.00-3.00
❏	❏	Bel Wa9503 **Squirt Gun - Shark in Blue** .	$2.00-3.00
❏	❏	Bel Wa9504 **Squirt Gun - Dolphin in Gray**.	$2.00-3.00

Comments: Distribution: Belgium - June, 1995; Germany - July, 1995; Parts of Europe, Middle East, and North Africa in 1995.

Fra Wa9564. French Translite.

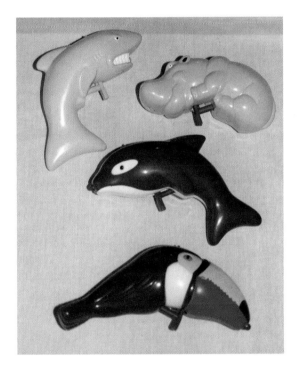

Bel Wa9501-04

Winnie the Pooh Tumblers Happy Meal, 1995

❏	❏	Zea Wi9501 **Tumbler: Pooh and the Honey Tree - yellow top & straw.** $4.00-5.00	
❏	❏	Zea Wi9502 **Tumbler: Pooh and Tigger Too - orange top & straw.** $4.00-5.00	
❏	❏	Zea Wi9503 **Tumbler: Pooh and Blustery Day - purple top & straw.** $4.00-5.00	
❏	❏	Zea Wi9504 **Tumbler: Pooh and Day for Eeyore - blue top & straw.** $4.00-5.00	

Comments: Distribution: New Zealand- 1995.

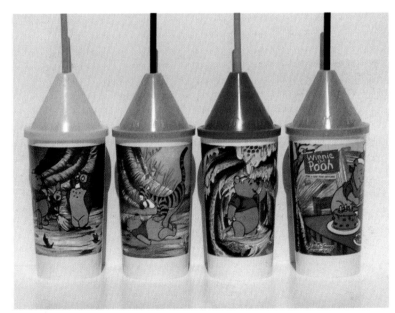

Zea Wi9501-04

Ger Wa9564. German Translite.

Winter Sports Happy Meal, 1995

❑ ❑	Zea Wi9501	**Grimace in Green Snow Tractor.**	
			$2.00-4.00
❑ ❑	Zea Wi9502	**Ronald on Skis.**	$2.00-4.00
❑ ❑	Zea Wi9503	**Birdie Skating.**	$2.00-4.00
❑ ❑	Zea Wi9504	**Hamburglar in Snow Mobile.**	$2.00-4.00

Comments: Distribution: New Zealand - 1995; France, Germany - 1995; UK - 1994 and 1995; parts of Europe in 1995.

Zea Wi9501-04

Japan Winter Sports

Ger Wi9526. German Display Danglers.

Fra Wi9564. French Translite for Winter Sports.

1996

- 13th National O/O Convention

- "QSC and Me" - Owner/Operator advertising theme

Aladdin and the King of Thieves Happy Meal, 1996

Premiums: Figures w Dioramas
- ❏ ❏ USA Al9601 **#1 Cassim W Diorama** - Standing W Dark Blue Cape/Black Wrap/Grey Boots/2p. $2.00-2.50
- ❏ ❏ USA Al9602 **#2 Abu W Diorama** - Monkey on Carpet W Backdrop/2p. $2.00-2.50
- ❏ ❏ USA Al9603 **#3 Jasmine W Diorama** - w Wht Dress/Gold Trim/Diorama/2p. $2.00-2.50
- ❏ ❏ USA Al9604 **#4 Lago W Diorama** - Red Bird on Blue Money Cart/Diorama/2p. $2.00-2.50
- ❏ ❏ USA Al9605 **#5 Genie W Diorama** - Blue Genie on Wht Cloud/2p. $2.00-2.50
- ❏ ❏ USA Al9606 **#6 Sa'luk W Diorama** - Blk/Purp Bare Chested/Gold Wrist Band. $2.00-2.50
- ❏ ❏ USA Al9607 **#7 Aladdin W Diorama** - Wht Coat/Gold Trim/ Hand to Waist/2p. $2.00-2.50
- ❏ ❏ USA Al9608 **#8 Mai Tre D'genie** - Purp Fig/Blk Long Coat/ Wht Wedding Cake/2p. $2.00-2.50

Comments: Distribution: USA - August 16-September 12, 1996; Germany (#2, 3, 5,7) - January, 1997.

USA Al9601-03

USA Al9604-06

USA Al9607-08

Ger Al9664. German Translite.

Babe Happy Meal, 1996

Premiums: Stuffed Barnyard Animals
- ❏ ❏ USA Bb9601 **#1 Babe the Pig** - Stuffed/Plush Pig/Pink . $2.00-2.50
- ❏ ❏ USA Bb9602 **#2 Cow** - Stuffed/Plush Cow/Blk-Wht. $2.00-2.50
- ❏ ❏ USA Bb9603 **#3 Maa the Ewe** - Stuffed/Plush Sheep/Wht. $2.00-2.50
- ❏ ❏ USA Bb9604 **#4 Fly the Dog** - Stuffed/Plush Dog/Blk-Wht. $2.00-2.50
- ❏ ❏ USA Bb9605 **#5 Ferdinand the Duck** - Stuffed/Plush Wht Duck. $2.00-2.50
- ❏ ❏ USA Bb9606 **#6 Dutchess the Cat** - Stuffed/Plush Furry Grey Cat. $2.00-2.50
- ❏ ❏ USA Bb9607 **#7 Mouse** - Stuffed Mouse/Plush Sitting Up- right. $2.00-2.50

Comments: National Distribution: USA - June 14 - July 11, 1996.

USA Bb9601-04

USA Bb9605-07

Comments: National Distribution: USA - July 12-August 15, 1996. **During clean-up weeks in England a few stores distributed Barbie Variation dolls: Weekend Barbie (same as USA All American, but no "Reebock" on shoe) and Wedding Barbie (same as USA Wedding Day Midge Barbie, but with blonde hair.) Also, Canada and New Zealand distributed variations: Birthday Party Barbie with white/blonde hair and Paint N Dazzle Barbie with brown hair.** These variations have been distributed around the world, in other countries, during clean-up periods.

Barbie Dolls of the World/Hot Wheels VII Happy Meal, 1996

- ❏ ❏ USA Ba9601 **#1 Dutch Barbie** - Blue Wht Striped Skirt/Wht Hat/Blonde Braids. $2.00-4.00
- ❏ ❏ USA Ba9602 **#2 Kenyan Barbie** - Red Fabric Cape/Yel Base/Red Plastic Dress/2p. $2.00-4.00
- ❏ ❏ USA Ba9603 **#3 Japanese Barbie** - Lt Purp Fabric Kimona/Blk Hair. $2.00-4.00
- ❏ ❏ USA Ba9604 **#4 Mexican Barbie** - Wht/Red/Green Flowered Fabric Dress. $2.00 - 4.00
- ❏ ❏ USA Ba9605 **#5 U.S.A. Olympic Barbie** - Red/Wht Gymnastic Outfit **W Blue Tights, medals.** $3.00-5.00

- ❏ ❏ Can Ba9605 **#5 Canadian Barbie** - Red/Wht Gymnastic Outfit **W Red Tights, no medals.** $4.00-5.00

USA Hw9606-10

USA Ba9601-05

- ❏ ❏ USA Hw9606 **#6 Flame Series** - Metallic Blue W Org Flames. $2.50-3.00
- ❏ ❏ USA Hw9607 **#7 Roaring Rod Series** - Extended Length/Wht W Blk Zebra Stripes. $2.50-3.00
- ❏ ❏ USA Hw9608 **#8 Dark Rider Series** - Black/Batman Style Car. $2.50-3.00
- ❏ ❏ USA Hw9609 **#9 Hot Hubs Series** - Org/Exposed Silver Engine. $2.50-3.00
- ❏ ❏ USA Hw9610 **#10 Krackel Car Series** - Bright Grn W Yel Details. $2.50-3.00

French Poster.

Batman and Robin/Batman Bike Happy Meal, 1996

☑	❏	Eur Bt9601	**Batman Mirror**.	$4.00-5.00
❏	❏	Eur Bt9602	**Batmobile Drink Bottle**.	$4.00-5.00
❏	❏	Eur Bt9603	**Batman Toolbox**.	$4.00-5.00
☑	❏	Eur Bt9604	**Joker's Horn**.	$4.00-5.00

Comments: Distribution: Europe - May/June 1996.

Eur Bt9601-04

Fra Bt9626. French Display.

Fra Bt9664. French Translite.

Carnival Thrills/Carnival Ride Happy Meal, 1996

❏	❏	Zea Ca9601 **Airplane Ride** - Green Base With Hamburglar on Orange Planes. $3.00-4.00
❏	❏	Zea Ca9602 **Carousel** - Dark Purple Base, Orange/Reddish Middle With Yellow Top. $3.00-4.00
❏	❏	Zea Ca9603 **Ferris Wheel** - Yellow Wheel With Purple Base. $3.00-4.00
❏	❏	Zea Ca9604 **Rocket Ride** - Yellow Scoop Rides on Purple & Pink Base. $3.00-4.00

Comments: Distribution: New Zealand - 1996. Each set came with a sticker sheet and toys marked, "China 1996."

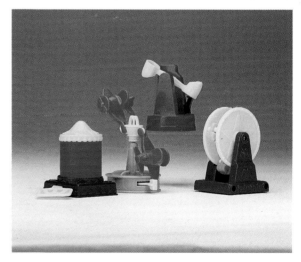

Zea Ca9601-04

1996

Crazy Feet Cup Happy Meal, 1996

☐ ☐ Aus Cr9601 **Yellow Cup/Red Feet - Ronald**. $2.00-4.00
☐ ☐ Aus Cr9602 **Purple Cup/Purple Feet - Grimace**.
$2.00-4.00
☐ ☐ Aus Cr9603 **Pink Cup/Yellow Feet - Birdie**. $2.00-4.00
☐ ☐ Aus Cr9604 **Orange Cup/White Feet - Hamburglar**.
$2.00-4.00

Comments: Distribution Australia - July/August 1996.

Aus Cr9601-04

Dingo Et Max/Goofy Obercool/Groovy Movie Happy Meal, 1996

☐ ☐ Eur Go9601 **Goofy Max with Glasses**. $2.00-4.00
☐ ☐ Eur Go9602 **Ape Big Foot - Large**. $2.00-4.00
☐ ☐ Eur Go9603 **Girl Roxanne**. $2.00-4.00
☐ ☐ Eur Go9604 **Goofy Dingo W Fishing Rod**. $2.00-4.00

Comments: Distribution: Europe - 1996; Germany - August, 1996;
France - June/July 1996.

Eur Go9601-04

German Translite.

French Translite.

Dinosaurs/Prehisteria Happy Meal, 1996

☐ ☐ Chi Di9601 **Dinosaur Skeleton -** Blue/3p. $2.00-3.00
☐ ☐ Chi Di9602 **Dinosaur Skeleton -** Green/3p. $2.00-3.00
☐ ☐ Chi Di9603 **Dinosaur Skeleton -** Purple/3p. $2.00-3.00
☐ ☐ Chi Di9604 **Dinosaur Skeleton -** Yellow/3p. $2.00-3.00

Comments: Distribution: China, Argentina, Brazil - September,
1996. Called Prehisteria in South America. The two piece plastic dinosaurs pull apart to reveal plastic skeletons of dinosaurs.

Chi Di9601-04

Disneyland Paris Happy Meal, 1996

- 🔲 ❏ Fra Di9601 **Mickey in Castle** - pink bottom, blue top.
  $4.00-5.00
- 🔲 ❏ Fra Di9602 **Daisy Duck in Small World -** pink bottom and top.
  $4.00-5.00
- 🔲 ❏ Fra Di9603 **Minnie in House** - purple or pink bottom and top.
  $4.00-5.00
- ❏ ❏ Fra Di9604 **Donald in Space Mountain** - blue bottom, red top.
  $4.00-5.00
- ❏ ❏ Jpn Di9605 **Mickey in Castle** - beige bottom, blue top.
  $4.00-5.00
- ❏ ❏ Jpn Di9606 **Donald in Space Mountain** - green bottom, red top.
  $4.00-5.00

Comments: Distribution: Europe - January/February 1996. Several countries varied the colors of the tops and bottoms of the Disney houses and cottage holding the Disney characters.

Jpn Di9605-06, Fra Di9602-03

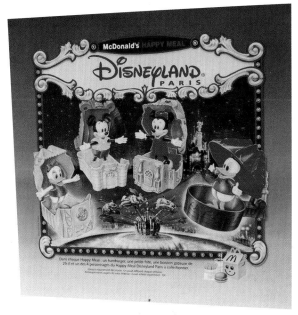

French Translite.

Ger Di9626. German Display Danglers.

Fra Di9601-04

Fra Di9604

Ger Di9601. German Display Dangler.

Dragon Ball Happy Meal, 1996

❏ ❏ Jpn Dr9601 **Picture Puzzle.** $3.00-4.00
❏ ❏ Jpn Dr9602 **Note Pad.** $3.00-4.00
❏ ❏ Jpn Dr9603 **Action Pictures.** $3.00-4.00
❏ ❏ Jpn Dr9604 **Pogs - 3 large.** $3.00-4.00

Comments: Distribution: Japan - 1996.

Jpn Dr9601-04

Japanese Counter Card.

Eric Carle Finger Puppet Happy Meal, 1996

❏ ❏ USA Er9601 **#1 The Very Quiet Cricket -** Dk Blue Head W Purple Body/Clip Button on Bottom. $2.00-2.50
❏ ❏ USA Er9602 **#2 The Grouchy Ladybug -** Red Wings/Peg on Bottom Moves Wings. $2.00-2.50
❏ ❏ USA Er9603 **#3 The Very Busy Spider -** Wht Base/Finger Moves Head & Body. $2.00-2.50
❏ ❏ USA Er9604 **#4 The Very Hungry Caterpillar -** Red Apple/ Finger Moves Worm's Head & Body. $2.00-2.50
❏ ❏ USA Er9605 **#5 A House for Hermit Crab -** Wht Snail Shell/Finger Moves Crab's Body. $2.00-2.50
❏ ❏ USA Er9606 **#6 The Very Lonely Firefly -** Red Wings/Grn Head/Finger Moves Firefly. $2.00-2.50

Comments: National Distribution: USA - September 20-October 10, 1996.

USA Er9601-06

Fisher-Price Under-3 Promotion, 1996

Group A Premiums:
❏ ❏ USA Fp9601 **Ronald at the Drive-Thru -** Yellow Building/ Red Roof. $3.00-4.00
❏ ❏ USA Fp9602 **Grimace Inside Purple Rolling Ball.** $3.00-4.00
❏ ❏ USA Fp9603 **Birdie in Aqua Pop Car W Pink Wheels.** $3.00-4.00
❏ ❏ USA Fp9604 **Yellow/Red "M" Shaped Key With White Ring.** $3.00-4.00

USA Fp9601-04

Group B Premiums:

❏ ❏ USA Fp9605 **White Train Engine W Blue Wheels -** Red/
Yellow Spinner on Top. $3.00-4.00

❏ ❏ USA Fp9606 **Barn Puzzle Square -** Yellow and Red sides.
                                          $3.00-4.00

❏ ❏ USA Fp9607 **Boom Box Radio -** Yellow Front/Green Back.
                                          $3.00-4.00

❏ ❏ USA Fp9608 **Peg in a Barrel -** White or Red side/House
Shaped. $3.00-4.00

Group D Premiums:

❏ ❏ USA Fp9613 **Chatter Telephone/Wheels -** White Phone/
Yellow Dial/Red Receiver. $3.00-4.00

❏ ❏ USA Fp9614 **Rolling Blue Ball W Dog Inside**.
                                          $3.00-4.00

❏ ❏ USA Fp9615 **Red Clock -** Blue Roof/Yellow Hands.
                                          $3.00-4.00

❏ ❏ USA Fp9616 **Musical Ball -** Yellow/Blue/Red. $3.00-4.00

USA Fp9605-08

USA Fp9613-16

Group C Premiums:

❏ ❏ USA Fp9609 **Doghouse with Puppy inside Window -** Red
Roof/Blue Front. $3.00-4.00

❏ ❏ USA Fp9610 **Yellow Round Open Ball/Yellow/Red/Blue/
Twirler -** Hands Grasp Ball Shape. $3.00-4.00

❏ ❏ USA Fp9611 **Lime Green Jeep with Purple Wheels -** No
Driver, No Passengers, Blue Seats. $3.00-4.00

❏ ❏ USA Fp9612 **Book on Cows -** Washable Plastic Type/Red/
White/Yellow/Blue. $3.00-4.00

Group E Premiums:

❏ ❏ USA Fp9617 **Blue/Red Roof Dog House W Brn Eye Dog
Inside**. $3.00-4.00

❏ ❏ USA Fp9618 **White Dog W Brown Spots on Red Wheels**.
                                          $3.00-4.00

❏ ❏ USA Fp9619 **Poppity Red Car W Yellow Wheels -** Red
Car W Yellow Wheels/People Inside. $3.00-4.00

❏ ❏ USA Fp9620 **McDonald's Truck -** Red Cab/White Truck Bed/
Blue Wheels. $3.00-4.00

USA Fp9617-20

USA Fp9612, 9609

USA Fp9610-11

Group F Premiums

❏ ❏ USA Fp9621 **Balls in a Yellow Ball -** Yellow Ball W Red &
White Balls Inside. $3.00-4.00

❏ ❏ USA Fp9622 **School Bus -** Yellow Bus W Blue Wheels/Ronald
McDonald Driving. $3.00-4.00

❏ ❏ USA Fp9623 **Horse -** White W Red Reins. $3.00-4.00

❏ ❏ USA Fp9624 **Lawn Mower Popper -**Push Long Handle/
Items Inside Pop Up And Down. $3.00-4.00

USA Fp9621-24

Comments: National Distribution: USA - September, 1996-August, 1997. Several countries around the world distributed these toys during clean-up periods. McDonald's standardization move resulted in a generic type Fisher-Price U-3 toy distributed for children under the age of 3 in the USA, Canada, and Mexico. A combination of 8, 12 or more of the 24 announced Fisher-Price U-3 Toys were distributed. Variations of the above toys exist with the molds being made backwards/reversed. That is, an "M" key on a ring exists with a red front/yellow back and white ring at top instead of USA Fp9604. For identification purposes, the back of the toy is considered where the screws can be viewed. This mold variation exists for several of the U-3 toys. The primary basic toy is listed above; prices on the various toys and/or variations are the same. The new package used is red, white yellow, and blue. The front of the package has Ronald McDonald holding up three fingers to indicate the age: 3. The U-3 poster displayed at some stores does not include USA Fp9624, the Lawn Mower Popper. This U-3 was initially recalled due to misspelling of "Fishr-Price" instead of: "Fisher-Price" on the reverse side. The toy was reprinted with the correct spelling and distributed in the last box, Group F premiums.

USA Fp9607. Variation.

USA Fp9620

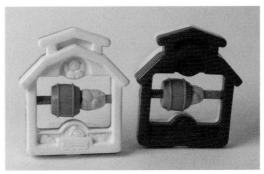

USA Fp9608. Variation.

USA Fp9604. Variation.

Fun Balls Happy Meal, 1996

- ❑ ❑   Aus Fb9601 **Rubber Ball - Birdie in pink.**
  $2.00-3.00
- ❑ ❑   Aus Fb9602 **Rubber Ball - Hamburglar in orange.**
  $2.00-3.00
- ❑ ❑   Aus Fb9603 **Rubber Ball - Ronald in yellow.**
  $2.00-3.00
- ❑ ❑   Aus Fb9604 **Rubber Ball - Grimace in purple.**
  $2.00-3.00

Comments: Distribution: Australia - October/November 1996.

USA Fp9606. Variation.

Aus Fb9601-04

Generic Happy Meal Premiums, 1996

❏ ❏ Aus Ge9601 **Spoon** - Ronald in Yellow. $1.00-2.00
❏ ❏ Aus Ge9602 **Fork** - Birdie in Pink. $1.00-2.00
❏ ❏ Aus Ge9603 **Comb** - Grimace in Purple. $1.00-2.00

Comments: Distribution: Australia 1996 during clean-up periods.

USA Ha9604-06

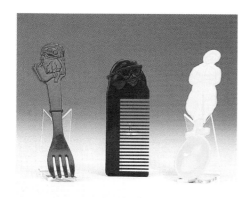

Aus Ge9601-03

Halloween '96/Monster McNuggets/McNugget Buddies Happy Meal, 1996

❏ ❏ USA Ha9601 **#1 Spider McNugget -** Purp Spider Legs Base W "M"/Purple Hat/Purp Eyelashes McNugget/3p.
  $1.00-1.50

❏ ❏ USA Ha9602 **#2 Rock Star McNugget -** Grn Florescent Hat/Blk Shoes Base W Yel "M"/Blk Eyelashes/3p.
  $1.00-1.50

❏ ❏ USA Ha9603 **#3 Fairy Princess McNugget -** Yel Hair W Silver Crown W "M"/Metallic Grn Base/Blu Eyelashes/3p.
  $1.00-1.50

❏ ❏ USA Ha9604 **#4 Dragon McNugget -** Grn Dragon Hat/ Grn Dragon Feet W Yel "M"/Red Mouth McNugget/3p .
  $1.00-1.50

❏ ❏ USA Ha9605 **#5 Alien Monster McNugget -** Org Hat/Pnk W 4 Arms Base/2 Teeth McNugget/3p. $1.00-1.50

❏ ❏ USA Ha9606 **#6 Ronald McNugget -** Red Hair/Yel Clown Base W Red Shoes/Red Smile McNugget/3p . $2.00-2.50

❏ ❏ USA Ha9627 **Fisher-Price Dragon/Knight -** Black Knight/ Red Florescent Dragon/2p. $4.00-5.00

USA Ha9627

Comments: National Distribution: USA - October 11-31, 1996. In some countries, this is a repeat of Halloween Happy Meal, 1994 from the USA concept.

USA Ha9601-03

Ger Ha9626. German Display Dangler.

49

Hunchback of Notre Dame Happy Meal, 1996

- Fra Hu9601 **Quasimodo Climbing** - With Pull String.
  $3.00-4.00
- Fra Hu9602 **Gargoyles (Hugo, Laverne, Victor) on gray stand** - Three Gargoyles With Cardboard Backdrop.
  $3.00-4.00
- Fra Hu9603 **Esmeralda on Stand** - Lady On Stand.
  $3.00-4.00
- Fra Hu9604 **Phoebus -** With Moveable Arms, Head, and Waist.
  $3.00-4.00
- Fra Hu9605 **Judge Frollo** - With Hat & Black Cloth Cape.
  $3.00-4.00
- Fra Hu9606 **Goat**.
  $3.00-4.00

Comments: Distribution: New Zealand - 1996; Europe - July/August, 1996; Germany - December, 1996, France, 1996. In some markets, the additional premium of a goat was distributed. Different numbers of premiums were distributed around the world, in an attempt to go global with this promotion. The Disney movie was released globally.

Ger Hu9626. German Display Dangler.

Zea Hu9601-05

German Translite.

Fra Hu9601-06

French Translite.

Hunchback of Notre Dame Lenticular Frames Happy Meal, 1996

❑ ❑   Aus Fr9601 **Dimensional Frame: Esmeralda**.
$2.00-4.00

❑ ❑   Aus Fr9602 **Dimensional Frame: Quasimodo**.
$2.00-4.00

❑ ❑   Aus Fr9603 **Dimensional Frame: Phoebus**. $2.00-4.00

❑ ❑   Aus Fr9604 **Dimensional Frame: Frollo**. $2.00-4.00

Comments: Distribution Australia - August/September, 1996. Each frame comes with three pictures that can be interchanged when the frame is opened up. Each picture is lenticular, one of two scenes can be viewed from different angles.

Aus Fr9601-04. Insert Card.

Island Getaway/McIsland/McHoliday/Pleasure Island/Summer Holiday Happy Meal, 1996

❑ ❑   Zea Is9601 **Birdie with a Large Fish on Line** - Whale on String. $3.00-4.00

❑ ❑   Zea Is9602 **Grimace Hula Dancing** - in Yellow Hula Skirt.
$3.00-4.00

❑ ❑   Zea Is9603 **Ronald in Hammock** - Under Palm Tree.
$3.00-4.00

❑ ❑   Zea Is9604 **Fry Guys Inside Orange Tiki Pole**.
$3.00-4.00

Comments: Distribution: New Zealand - 1996; Europe - 1996; France - July/August, 1996.

Zea Is9601-04

Ger Is9626. German Display Dangler.

Fra Is9626. French Display.

Fra Is9664. French Translite.

French Display.

Japanese Counter Card.

Lady and the Tramp/La Belle et le Clochard Happy Meal, 1996

☑	☐	Fra La9601	**Lady - Dog ina Doghouse**.	$4.00-5.00
☐	☐	Fra La9602	**Tramp - Dog in a Barrel**.	$4.00-5.00
☑	☐	Fra La9603	**Siamese Cats - Cats**.	$4.00-5.00
☑	☐	Fra La9604	**Jock - Dog With Long Ears**.	$4.00-5.00

Comments: Distribution: France - 1996.

French Translite.

Fra La9601-04

Lion King Happy Meal, 1996

❑ ❑   Aus Li9601 **Puzzle -** Timon and Pumbaa. $4.00-5.00
❑ ❑   Aus Li9602 **Crayon Board -** Timon and Pumbaa.
  $4.00-5.00
❑ ❑   Aus Li9603 **Mix N Match -** Timon and Pumbaa.
  $4.00-5.00
❑ ❑   Aus Li9604 **Sticky Picky Pad -** Timon and Pumbaa.
  $4.00-5.00

Comments: Distribution: Australia - 1996.

Aus Li9601-04

Littlest Pet Shop/Transformers Beast Wars I Happy Meal, 1996

Premiums:
❑ ❑   USA Li9601 **#1 Swan -** Pink/Purp Swan W Wings.
  $1.00-1.50
❑ ❑   USA Li9602 **#2 Unicorn -** Purple Unicorn W Pink Tail.
  $1.00-1.50
❑ ❑   USA Li9603 **#3 Dragon -** Yel Dragon W Org Wings.
  $1.00-1.50
❑ ❑   USA Li9604 **#4 Tiger -** White Tiger W Pink Fur.
  $1.00-1.50

❑ ❑   USA Li9605 **#5 Manta Ray -** Dk Blue/Lt Blu Ray W Movable Head. $1.00-1.50
❑ ❑   USA Li9606 **#6 Beetle -** Maroon/Grn/Grn Beetle W Movable Head. $1.00-1.50
❑ ❑   USA Li9607 **#7 Panther -** Blk/Blu/Red Panther W Movable Head. $1.00-1.50
❑ ❑   USA Li9608 **#8 Rhino -** Grey/Red Rhino W Movable Head . $1.00-1.50

Comments: Distribution: USA - March 15-April 11, 1996; Latin America - June, 1996.

USA Li9601-04

USA Li9605-08

Looney Tunes Parade Happy Meal, 1996

🖼 ❑   Ger Lo9601 **Bugs Bunny -** Sitting Playing Cymbals.
  $3.00-5.00
🖼 ❑   Ger Lo9602 **Tweety Bird -** in a Dryer Box. $3.00-5.00
🖼 ❑   Ger Lo9603 **Bull Dog/Taz -** Box Sides Open. $3.00-5.00
🖼 ❑   Ger Lo9604 **Roadrunner & Coyote/Carl -** Rock Back & Forth. $3.00-5.00

Comments: Distribution: Germany - July, 1996; France - December/January, 1996

Ger Lo9601-04

Ger Lo9626. German Display Dangler.

Fra Lo9664. French Translite.

Fra Lo9626. French Display.

Marvel Super Heroes Happy Meal, 1996

❑ ❑ USA Ma9601 **#1 Spider-Man Vehicle -** Car W Open/Close Webb Throw . $1.00-1.50

❑ ❑ USA Ma9602 **#2 Storm -** Storm in Parking Cloud. $1.00-1.50

❑ ❑ USA Ma9603 **#3 Wolverine -** Wolverine in a Jet W Retractable Claws. $1.00-1.50

❑ ❑ USA Ma9604 **#4 Jubilee -** Jubilee on a Scooter W Optical Illusion Shield. $1.00-1.50

❑ ❑ USA Ma9605 **#5 Invisible Woman -** Woman Changes Colors in Cold Water. $1.50-2.00

❑ ❑ USA Ma9606 **#6 Thing -** Vehicle Bursts Open to Reveal Thing. $1.00-1.50

❑ ❑ USA Ma9607 **#7 Hulk -** Hulk Figurine. $1.50-2.00

❑ ❑ USA Ma9608 **#8 Human Torch -** Human Torch Figurine . $1.00-1.50

Comments: National Distribution: USA - May 17-June 13, 1996.

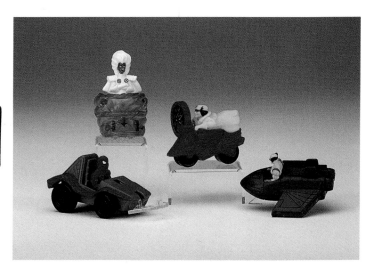

USA Ma9601-04

USA Ma9605-08

McFarm/a la Ferme/McGranja Happy Meal, 1996

❏ ❏   Zea Fa9601 **Birdie Pushing Chickens in a Green Basket**.
$3.00-4.00

❏ ❏   Zea Fa9602 **Grimace in Orange Popping Corn Car**.
$3.00-4.00

❏ ❏   Zea Fa9603 **Ronald in Green Tractor**. $3.00-4.00

❏ ❏   Zea Fa9604 **Hamburglar in Red Harvester**. $3.00-4.00

Comments: Distribution: New Zealand - 1996; Germany - March, 1996.

Zea Fa9601-04

Mexican Translite.

Ger Fa9626. German Display Dangler.

McSpace/McSpaceship Happy Meal, 1996

❏ ❏   Zea Sp9601 **#1 Ronald's Astro Repair Capsule**/3p.
$6.00-10.00

❏ ❏   Zea Sp9602 **#2 Grimace's Explorer Spacesuit Capsule**/
3p. $6.00-10.00

❏ ❏   Zea Sp9603 **#3 Hamburglar and his Alien Friends Capsule**/3p. $6.00-10.00

❏ ❏   Zea Sp9604 **#4 Birdie's Moon Buggy Capsule**/3p.
$6.00-10.00

Comments: Distribution: New Zealand - 1996. Four pieces connect together to form a silver spaceship. Two or more sizes of this spaceship exist—from approximately 9" to 13" in height when all sections are assembled. Around the world production of these toys vary, based on manufacturing plant.

Zea Sp9601-04

Zea Sp9601-04. Collect and Build Concept.

Muppet Treasure Island Color Me Kits Happy Meal, 1996

❏ ❏ Aus Mu9601 **Transfer with Yellow Box of Crayons.**
$2.00-3.00

❏ ❏ Aus Mu9602 **Transfer with Red Box of Crayons.**
$2.00-3.00

❏ ❏ Aus Mu9603 **Transfer with Purple Box of Crayons.**
$2.00-3.00

❏ ❏ Aus Mu9604 **Transfer with Blue Box of Crayons.**
$2.00-3.00

Comments: Distribution: Australia - June 21/July, 1996. Each transfer matched the design on the box of crayons, measured 6" x 8", and came with a box of crayons. The Happy Meal box formed a pirate's ship when extended.

Muppet Treasure Island Happy Meal, 1996

❏ ❏ USA Mu9601 **#1 Miss Piggy Tub Toy -** Miss Piggy in Grn Tub/Dress Changes Color. $1.00-1.25

❏ ❏ USA Mu9602 **#2 Kermit in Boat -** Kermit W Blk Cannon/ Brn Boat/Squirts Water. $1.00-1.25

❏ ❏ USA Mu9603 **#3 Gonzo in Paddle Wheel Boat -** Gonzo in Purp Boat W Grn Moving Wheel. $1.00-1.25

❏ ❏ USA Mu9604 **#4 Fozzie in Barrel -** Fossie in Blue Barrel/ Bobs up and Down. $1.00-1.25

Comments: Distribution: USA - February 16-March 14, 1996; New Zealand - 1996.

USA Mu9601-04

Musical Instruments Happy Meal, 1996

❏ ❏ Eur Mu9601 **Ronald Whistle.** $2.00-3.00
❏ ❏ Eur Mu9602 **Birdie Harmonica.** $2.00-3.00
❏ ❏ Eur Mu9603 **Hamburglar Twirl Noisemaker.**
$2.00-3.00
❏ ❏ Eur Mu9604 **Grimace Goose Shaped Tea Pot Noise-maker.** $2.00-3.00

Comments: Distribution: Europe - 1996 during clean-up periods; USA - 1997 during clean-up periods.

Eur Mu9601-04

Olympic Maze Puzzles Happy Meal, 1996

❏ ❏ Aus Ol9601 **Ronald Maze.** $2.00-4.00
❏ ❏ Aus Ol9602 **Birdie Maze.** $2.00-4.00
❏ ❏ Aus Ol9603 **Grimace Maze.** $2.00-4.00
❏ ❏ Aus Ol9604 **Hamburglar Maze.** $2.00-4.00

Comments: Distribution: Australia - April/May, 1996.

Olympic Pinball Games Happy Meal, 1996

❏ ❏ Aus Ol9601 **Pinball Game -** Ronald, orange, rectangle.
$4.00-5.00

❏ ❏ Aus Ol9602 **Pinball Game -** Grimace, purple, square.
$4.00-5.00

❏ ❏ Aus Ol9603 **Pinball Game -** Birdie, pink, triangle.
$4.00-5.00

❏ ❏ Aus Ol9604 **Pinball Game -** Hamburglar, yellow, half oval.
$4.00-5.00

Comments: Distribution: Australia - 1996.

Olympic/World Cup Tumblers Happy Meal, 1996

❏ ❏ Zea Tu9601 **Cup: Birdie Gymnastics -** purple lid with green straw. $2.00-3.00

❏ ❏ Zea Tu9602 **Cup: Grimace Running -** teal green lid with purple straw. $2.00-3.00

❏ ❏ Zea Tu9603 **Cup: Hamburglar Cycling -** orange lid with blue straw. $2.00-3.00

❏ ❏ Zea Tu9604 **Cup: Ronald Swiming -** blue lid with yellow straw. $2.00-3.00

❏ ❏ *** Tu9605 **Cup: Izzy with Basketball.** $2.00-3.00
❏ ❏ *** Tu9606 **Cup: Izzy with Soccerball.** $2.00-3.00
❏ ❏ *** Tu9607 **Cup: Izzy with Torch.** $2.00-3.00
❏ ❏ *** Tu9608 **Cup: Izzy with Hurdles.** $2.00-3.00

❏ ❏ Zea Tu9609 **Cup: Ronald Swimming.** $2.00-3.00
❏ ❏ Zea Tu9610 **Cup: Grimace Racing.** $2.00-3.00
❏ ❏ Zea Tu9611 **Cup: Hamburglar Driving.** $2.00-3.00
❏ ❏ Zea Tu9612 **Cup: Birdie with Streamers.** $2.00-3.00

Comments: Distribution: New Zealand - 1996. Multiple countries around the world distributed cups to celebrate special events, included within the Happy Meal purchase.

Riad, Saudia Arabia

Zea Tu9609-12

101 One Hundred and One Dalmatians Happy Meal II, 1996

Row 1 - Poster

❏ ❏ USA On9601 **Gold Wrap -** Wrapped in Gold Ribbon. $3.00-4.00

❏ ❏ USA On9602 **Orange Collar -** Orange Collar/1 Ear Spotted. $3.00-4.00

❏ ❏ USA On9603 **In Blue Truck -** in Blue/Brn Truck/Hiding Eyes/ Light Green Collar. $3.00-4.00

❏ ❏ USA On9604 **Wearing Brown Derby Hat -** Sitting up/Green Collar/Brown Boller Hat W Blue Hat Band.

$4.00-5.00

USA On9601-04

*** Tu9605-08

1996

❏ ❏ USA On9605 **Toy Soldier in Mouth -** Org Collar W Toy Soldier Nutcracker/Soldier in Mouth. $3.00-4.00

❏ ❏ USA On9606 **Dk Blue Collar/1 Black ear -** Dark Blue Collar/1 All Black Ear/1 Spotted Ear. $3.00-4.00

❏ ❏ USA On9607 **Green Wreath on Tail -** Light Green Collar W Grn Wreath on Rump. $3.00-4.00

❏ ❏ USA On9608 **Holly flower on Tail/Pink Collar -** Standing/ Pink Collar/Green Holly W Yellow Jingle Bells on Tail. $3.00- 4.00

USA On9613-16

USA On9605-08

Row 3 - Poster

❏ ❏ USA On9617 **Plain Green collar -** Mouth open/Spotted Ears. $3.00-4.00

❏ ❏ USA On9618 **In Pink Baby Buggy -** W Turq Wheels/Blue Collar/Hiding Eyes. $5.00-8.00

❏ ❏ USA On9619 **In Newspaper -** in White London Herald Newspaper House. $7.00-10.00

❏ ❏ USA On9620 **Wrapped in Red Scarf W Yellow -** Sweater/ scarf w Yellow Tassels. $3.00-4.00

Row 2 - Poster

❏ ❏ USA On9609 **Pink/Fuchsia collar/2 Black ears -** Fuchsia Collar/2 Solid Black Ears. $3.00-4.00

❏ ❏ USA On9610 **Wearing Cowboy Hat -** Dark Green Collar/ Brn Cowboy Hat W Red Band. $5.00-8.00

❏ ❏ USA On9611 **In Pink Tea Pot.** $12.00-15.00

❏ ❏ USA On9612 **Blue Mitten in Mouth -** Yellow Collar. $3.00-4.00

❏ ❏ USA On9613 **Newspaper in Mouth -** Green Collar. $3.00-4.00

❏ ❏ USA On9614 **Wrapped in Yellow/Green Scarf/Sweater -** No Collar. $3.00-4.00

❏ ❏ USA On9615 **Brown Bucket on Tail -** Orange Collar. $3.00-4.00

❏ ❏ USA On9616 **Wearing Green Santa Hat/Holding Bone -** Standing/Blue Collar. $3.00-4.00

USA On9617-20

USA On9609-12

58

❏ ❏ USA On96**21 Christmas Lights Wrapped Around Face -** Blue Collar/Sitting/Red Christmas Lights Around Face.
$5.00-6.00

❏ ❏ USA On96**22 Dog on 101 Drum -** on Heavy Red 101 Dalmatian Drum/Purple Collar. $7.00-10.00

❏ ❏ USA On96**23 Plain Yellow Collar -** Yellow Collar/Spotted Ears. $3.00-4.00

❏ ❏ USA On96**24 Black Rings Around Eye -** Dark Purple Collar/Black Eye/2 Solid Black Ears. $3.00-4.00

❏ ❏ USA On96**25 In Red Present -** Orange Collar.$5.00-8.00

❏ ❏ USA On96**26 Candy Cane in Mouth -** Sitting W Red/White Candy Cane in Mouth//Green Collar. $3.00-4.00

❏ ❏ USA On96**27 Holding Red Book on Belly -** Holding Red Book/Green Collar/Black Eye. $3.00-4.00

❏ ❏ USA On96**28 In Yellow Cookie Jar -** Green Collar.
$8.00-15.00

USA On9629-32

❏ ❏ USA On96**36 Bluebird on Head -** Red-Orange Collar.
$5.00-8.00

❏ ❏ USA On96**37 Wrapped in Blue Scarf -** Purple Tassels.
$3.00-4.00

❏ ❏ USA On96**38 In White DeVil Car -** Orange Collar.
$7.00-10.00

❏ ❏ USA On96**39 Yellow Bow on Tail -** Standing/Purple Collar/One Black Foot. $3.00-4.00

❏ ❏ USA On96**40 In Green Wreaths W Red Bow -** Wearing Red Santa Hat/Blue Collar. $5.00-8.00

❏ ❏ USA On96**41 Holding Cookie -** Standing W Blue Collar.
$3.00-4.00

USA On9621-24

USA On9625-28

USA On9633-36

USA On9637-41

Row 4 - Poster

❏ ❏ USA On96**29 Brown Shoe in Muth -** Holding Brown Shoe in Mouth/Purple Collar/1 Solid Black Ear. $3.00-4.00

❏ ❏ USA On96**30 In Yellow Present -** Coming out of Yellow Present/Blue Collar/Red Santa Hat. $5.00-8.00

❏ ❏ USA On96**31 With Green Candy Cane in Mouth/Holding Bone -** Standing W Purp Collar. $3.00-4.00

❏ ❏ USA On96**32 Light Blue Collar/White Ear -** Right Paw up W 1 Spotted Ear. $3.00-4.00

❏ ❏ USA On96**33 In Red Stocking -** Grn Collar/White Bone in Mouth/Black Eye. $3.00-4.00

❏ ❏ USA On96**34 Wearing black Cruella Hat -** Sitting W Red Collar. $3.00-4.00

❏ ❏ USA On96**35 Plain Purple Collar/White ears -** Dark Purple Collar/1 White Ear. $5.00-8.00

Row 5 - Poster

❏ ❏ USA On96**42** **Tennis Shoe in Mouth -** Laying Flat/Black-White Tennis Shoe in Mouth/Blue Collar. $3.00-4.00

❏ ❏ USA On96**43** **On Red Ornament -** on Heavy Red Christmas Ball Decorated W Green Tree/Yellow Collar. $7.00-10.00

❏ ❏ USA On96**44** **Wearing Indiana Jones Hat -** W Dk Pink Collar. $5.00-8.00

❏ ❏ USA On96**45** **Candle in Mouth -** W Org Candle in Mouth/ Dark Purple Collar. $3.00-4.00

❏ ❏ USA On96**46** **Wrapped in Green Ribbon -** I Black Ear/No Collar. $3.00-4.00

❏ ❏ USA On96**47** **Teddy Bear on Stomach -** Laying W Yellow Teddy Bear on Stomach/Orange Collar. $8.00-15.00

❏ ❏ USA On96**48** **In Green Bus -** W Purple Collar. $4.00-7.00

❏ ❏ USA On96**49** **Wrapped in Silver Ribbon -** 2 Spots on One Ear/No Collar. $3.00-4.00

❏ ❏ USA On96**50** **In Lavender Book -** W Purple Book. $10.00-15.00

❏ ❏ USA On96**51** **Wearing Mickey Mouse Ears -** W "Fidget" Name Back of Hat/Lt Blu Collar. $10.00-15.00

❏ ❏ USA On96**52** **In Blue Baby Buggy -** Yellow Wheels/Red Collar. $5.00-8.00

USA On9650-52

Row 6 - Poster

❏ ❏ USA On96**53** **Wrapped in Aqua Garland -** W Purple Ornaments/Spotted Head/No Collar. $3.00-4.00

❏ ❏ USA On96**54** **Wrapped in Green Wreath W Yellow Ribbon -** W Candy Cane in Mouth. $4.00-5.00

❏ ❏ USA On96**55** **Wearing Blue Bobby Hat -** W Red Collar. $3.00-4.00

❏ ❏ USA On96**56** **Wearing Crown on Head -** W Lavender Collar. $5.00-8.00

❏ ❏ USA On96**57** **In Red Bus -** W Light Green Collar. $3.00-4.00

❏ ❏ USA On96**58** **In Bobby Hat -** W Red Collar. $5.00-8.00

❏ ❏ USA On96**59** **Yellow Trumpet on Stomach -** Laying Down/ Playing Yellow Trumpet/Horn/Holly on Horn/Red Collar. $3.00-4.00

❏ ❏ USA On96**60** **Wrapped in Red Garland -** W Green Ornaments/No Collar. $3.00-4.00

USA On9642-45

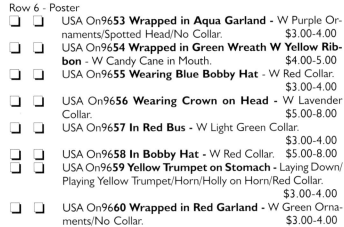

USA On9653-56

USA On9646-49

USA On9657-60

❏ ❏ USA On9661 **In Blue Present** - W Bone in Mouth/Silver Ribbon/Purple Collar. $5.00-8.00

❏ ❏ USA On96**62 Leash Wrapped Around Face -** W Green Collar. $5.00-6.00

❏ ❏ USA On96**63 In Silver Paint Can -** W Purple Paint. $7.00-10.00

❏ ❏ USA On96**64 Holding Bone in Mouth -** W Light Green Collar. $3.00-4.00

USA On9669-72

USA On9661-64

Row 7 - Poster

❏ ❏ USA On96**65 On Soccer Ball -** W Yellow Collar. $5.00-8.00

❏ ❏ USA On96**66 Solid Black Tail Painted On -** W Fuchsia/Red collar/No Spots on Ears. $3.00-4.00

❏ ❏ USA On96**67 Holding Dog Dish in Mouth -** W Green Collar. $3.00-4.00

❏ ❏ USA On96**68 Green Frog on Head -** W Pink Fuchsia Collar. $5.00-8.00

❏ ❏ USA On96**69 Blue Top Hat on Tail -** W Green Collar. $3.00-4.00

❏ ❏ USA On96**70 Butterfly on Head -** W Green/Teal Collar. $5.00-8.00

❏ ❏ USA On96**71 Red Flower/Bow on Nose -** W Yellow Collar. $3.00-4.00

❏ ❏ USA On96**72 Holding Blue Can of Dog Treats -** W Yellow Collar. $3.00-4.00

❏ ❏ USA On96**73 Red Bow on Tail -** W Blue Collar. $3.00-4.00

❏ ❏ USA On96**74 Holding Red Dog Dish -** W Purple Collar. $3.00-4.00

❏ ❏ USA On96**75 Stick in Mouth -** W Pinkish-Lavender Collar. $3.00-4.00

❏ ❏ USA On96**76 In Red Doghouse -** W Green Roof. $4.00-5.00

USA On9673-76

Row 8 - Poster

❏ ❏ USA On96**77 Holding Purple Present -** W Lime Green Collar. $3.00-4.00

❏ ❏ USA On96**78 In Green Stocking -** W Red-White Candy Cane in Mouth/Blue Collar. $3.00-4.00

❏ ❏ USA On96**79 Wearing Black Palace Hat -** W Purple Collar. $3.00-4.00

❏ ❏ USA On96**80 Leash in Mouth -** W Orange Collar. $3.00-4.00

USA On9677-80

USA On9665-68

❏ ❏ USA On96**81 Holding Green Bell -** W Red Holly/Fuchsia Collar. $3.00-4.00

❏ ❏ USA On96**82 Purple Bow on Head** - W Blue Collar. $5.00-6.00

❏ ❏ USA On96**83 In Brown Barrel** - W Blue Collar. $5.00-8.00

❏ ❏ USA On96**84 In Purple Umbrella -** W Green Collar. $15.00-20.00

❏ ❏ USA On96**85 Wearing Grey Wig -** W Fuchsia Collar. $5.00-6.00

❏ ❏ USA On96**86 On Green Turtle -** W Red-Orange Collar. $15.00-20.00

❏ ❏ USA On96**87 Pushing Blu/Wht Soccer Ball -** W Orange Collar. $3.00-4.00

❏ ❏ USA On96**88 Red Ornament on nose -** W Purple Collar. $3.00-4.00

❏ ❏ USA On96**89 Orange Maple Leaf on Head -** W Fuchsia Collar. $5.00-8.00

Row 9 - Poster

❏ ❏ USA On96**90 Wearing Green Santa Stocking -** W Red Collar. $3.00-4.00

❏ ❏ USA On96**91 Wearing Purple Beret** - W Yellow Collar. $5.00-8.00

❏ ❏ USA On96**92 In Black Rubber Tire -** W Red Collar. $7.00-10.00

❏ ❏ USA On96**93 Wearing Blue Baseball Cap -** W Orange Collar. $3.00-4.00

❏ ❏ USA On96**94 Gold Present in Mouth -** W Blue Collar. $3.00-4.00

❏ ❏ USA On96**95 Green Wreath around Nose -** W Yellow Collar. $3.00-4.00

❏ ❏ USA On96**96 Green Cricket on Head -** W Purple Collar. $5.00-8.00

❏ ❏ USA On96**97 Pink Present on Tail -** W Yellow Collar. $3y.00-4.00

❏ ❏ USA On96**98 Brown Purse in Mouth -** W Red Collar. $3.00-4.00

❏ ❏ USA On96**99 Red Santa Hat on Tail -** W Purple Collar. $3.00-4.00

❏ ❏ USA On96**100 Wearing Orange Hunting Cap -** W Lime Green Collar. $3.00-4.00

❏ ❏ USA On96**101 In Red Car -** W Purple Collar. $3.00-4.00

USA On9681-84

USA On9685-89

USA On9690-93

USA On9694-97

USA On9698-101

Variation:
❏ ❏   USA On96**102 Green Present in Mouth -** W Blue Collar.
         $20.00-25.00

Comments: National Distribution: USA/Canada/Mexico - November 27, 1996 - January 2, 1997; Germany (Full series of 101) - March 1997.

Ger On9626. German Display Dangler.

101 Dalmatians Alternate Checklist
(Arranged according to type of position and collar color with numbers relating to position on the posters)

Wearing Collars Only
❏	❏	# 2 Orange Collar.	$3.00-4.00
❏	❏	# 6 Dark Blue Collar W 1 All Black Ear.	$3.00-4.00
❏	❏	# 32 Light Blue Collar W 1 Spotted Ear.	$3.00-4.00
❏	❏	# 9 Light Pink/Fuchsia-Purplish Collar W 2 Blk Ears.	
			$3.00-4.00
❏	❏	# 66 Fuchsia-Red collar W Solid Painted on Black Tail.	
			$3.00-4.00
❏	❏	# 17 Green Collar.	$3.00-4.00
❏	❏	# 23 Yellow Collar.	$3.00-4.00
❏	❏	# 24 Dark Purple Collar W 1 Black Eye.	$3.00-4.00
❏	❏	# 35 Dark Purple Collar W 1 White Ear.	$3.00-4.00

Puppies in Vehicles
❏	❏	# 3 In Blue Truck, Light Green Collar.	$3.00-4.00
❏	❏	# 18 In Pink Doll Carriage, Blue Collar.	$5.00-8.00
❏	❏	# 38 In White Devil Car, Orange Collar.	$7.00-10.00
❏	❏	# 48 In Green Bus, Purple Collar.	$4.00-7.00
❏	❏	# 52 In Blue Doll Carriage, Red Collar.	$5.00-8.00
❏	❏	# 57 In Red Bus, Green Collar.	$3.00-4.00
❏	❏	# 101 In Red Car, Purple Collar.	$3.00-4.00

Wearing A Hat
❏	❏	# 4 Wearing Brown Bowler Hat, Green Collar.	$4.00-5.00
❏	❏	#10 Wearing Brown Cowboy Hat, Dark Green Collar.	
			$5.00-8.00
❏	❏	#35 Wearing Cruella's Black Hat, Red Collar.	$5.00-8.00
❏	❏	#44 Wearing Tan Indiana Jones Hat, Dark Pink Collar.	
			$5.00-8.00
❏	❏	#51 Wearing Mickey Mouse Ears, "Fidget" Name, Light Blue Collar.	$10.00-15.00
❏	❏	#55 Wearing Blue Bobby Hat, Red Collar	3.00-4.00
❏	❏	#56 Wearing Purple Crown, Purple Collar.	$5.00-8.00
❏	❏	#79 Wearing Black Bobby Hat, Purple Collar.	$3.00-4.00
❏	❏	#91 Wearing Purple Beret, Yellow Collar.	$5.00-8.00
❏	❏	#93 Wearing Blue Baseball Cap, Orange Collar.	
			$3.00-4.00
❏	❏	#100 Wearing Orange Hunting Hat, Light Green Collar.	
			$3.00-4.00

Holding Something
❏	❏	#16 Holding Bone, Wearing Green Santa Hat, Blue Collar .	
			$3.00-4.00
❏	❏	#27 Holding Red Book, Green Collar.	$3.00-4.00
❏	❏	#41 Holding Brown Cookie, Blue Collar.	$3.00-4.00
❏	❏	#47 Holding Yellow Bear, Orange Collar.	$8.00-15.00
❏	❏	#59 Holding Yellow Horn, Red Collar.	$3.00-4.00
❏	❏	#72 Holding Blue Can, Yellow Collar.	$3.00-4.00
❏	❏	#74 Holding Red Bowl, Purple Collar.	$3.00-4.00
❏	❏	#77 Holding Purple Present, Lime Green Collar.	
			$3.00-4.00
❏	❏	#81 Holding Green Bell, Fuchsia Collar.	$3.00-4.00
❏	❏	#87 Holding Blue & White Soccer Ball, Orange Collar.	
			$3.00-4.00

With Scarves and/or Ribbons
❏	❏	# 1 Wrapped in Gold Ribbon.	$3.00-4.00
❏	❏	#14 Wrapped in Yellow & Green Scarf.	
			$3.00-4.00
❏	❏	#20 Wrapped in Red Scarf.	$3.00-4.00
❏	❏	#37 Wrapped in Blue Scarf.	$3.00-4.00
❏	❏	#46 Wrapped in Green ribbon.	$3.00-4.00
❏	❏	#49 Wrapped in Silver ribbon.	$3.00-4.00
❏	❏	#53 Wrapped in Turquoise Garland W Purple Ornaments.	
			$3.00-4.00
❏	❏	#60 Wrapped in Red Garland W Green Ornaments.	
			$3.00-4.00

On Top of Something

- ❏ ❏ #22 On 101 Dalmatians Drum, Purple Collar. $7.00-10.00
- ❏ ❏ #43 On Red Ornament, Yellow Collar. $7.00-10.00
- ❏ ❏ #65 On Blue & White Soccer Ball, Yellow Collar. $5.00-8.00
- ❏ ❏ #84 In Purple Umbrella, Green Collar. $15.00-20.00
- ❏ ❏ #86 On Green Turtle, Red-Orange Collar. $15.00-20.00
- ❏ ❏ #92 In Black Tire, Red Collar. $7.00-10.00

Inside of Something

- ❏ ❏ #11 In Pink Teapot. $12.00-15.00
- ❏ ❏ #19 In London Herald Newspaper. $7.00-10.00
- ❏ ❏ #25 In Red Present, Orange Collar. $5.00-8.00
- ❏ ❏ #28 In Yellow Cookie Jar, Green Collar $8.00-15.00
- ❏ ❏ #30 In Yellow Present, Blue Collar. $5.00-8.00
- ❏ ❏ #33 In Red Stocking, Green Collar. $3.00-4.00
- ❏ ❏ #40 In Red Box, Blue Collar. $5.00-8.00
- ❏ ❏ #50 In 101 Lavender Book. $10.00-15.00
- ❏ ❏ #54 In Green Wreaths W Yellow Bow. $4.00-5.00
- ❏ ❏ #58 In Blue Bobby Hat, Red Collar. $5.00-8.00
- ❏ ❏ #61 In Blue Present, Purple Collar. $5.00-8.00
- ❏ ❏ #63 In Silver Paint Can, Purple Paint. $7.00-10.00
- ❏ ❏ #76 In Red Dog House. $4.00-5.00
- ❏ ❏ #78 In Green Stocking, Blue Collar. $3.00-4.00
- ❏ ❏ #83 In Brown Barrel, Blue Collar. $5.00-8.00

Something in the Mouth

- ❏ ❏ # 5 Nutcracker Soldier in Mouth, Orange Collar. $3.00-4.00
- ❏ ❏ # 8 Turquoise Leash in Mouth, Orange Collar. $3.00-4.00
- ❏ ❏ #12 Blue Mitten in Mouth, Yellow Collar. $3.00-4.00
- ❏ ❏ #13 Newspaper in Mouth, Green Collar. $3.00-4.00
- ❏ ❏ #26 Red & White Candycane in Mouth, Green Collar. $3.00-4.00
- ❏ ❏ #29 Brown shoe in Mouth, Purple Collar. $3.00-4.00
- ❏ ❏ #31 Green & White Candycane in Mouth, Purple Collar. $3.00-4.00
- ❏ ❏ #42 Black & White Tennis Shoe in Mouth, Blue Collar. $3.00-4.00
- ❏ ❏ #45 Yellow Candle in Mouth, Purple Collar. $3.00-4.00
- ❏ ❏ #64 Bone in Mouth, Green Collar. $3.00-4.00
- ❏ ❏ #67 Red Bowl in Mouth, Green Collar. $3.00-4.00
- ❏ ❏ #75 Brown Stick in Mouth, Light Purple Collar. $3.00-4.00
- ❏ ❏ #94 Gold Present in Mouth, Blue Collar. $3.00-4.00
- ❏ ❏ #98 Brown Purse in Mouth, Red Collar. $3.00-4.00
- ❏ ❏ #102 Green Present in Mouth, Yellow Collar (Variation). $20.00-25.00

Something on Nose or Face

- ❏ ❏ #21 String of Lights Around Nose, Blue Collar. $5.00-6.00
- ❏ ❏ #62 Brown Leash Around Face, Green Collar. $5.00-6.00
- ❏ ❏ #71 Red Bow on Nose, Lime Green/Yellow Collar. $3.00-4.00
- ❏ ❏ #88 Red Ornament on Nose, Purple Collar. $3.00-4.00
- ❏ ❏ #95 Wreath Around Nose, Yellow Collar. $3.00-4.00

Something on Tail or Bottom

- ❏ ❏ # 7 Green Wreath on Tail, Light Green Collar. $3.00-4.00
- ❏ ❏ # 8 Jingle Bells W Holly on Tail, Pink Collar. $3.00-4.00
- ❏ ❏ #15 Brown Bucket on Tail, Orange Collar. $3.00-4.00
- ❏ ❏ #39 Yellow Bow on Tail, Purple Collar. $3.00-4.00
- ❏ ❏ #69 Blue Top Hat on Tail, Green Collar. $3.00-4.00
- ❏ ❏ #73 Red Bow on Tail, Blue Collar. $3.00-4.00
- ❏ ❏ #90 Green Stocking on Tail, Red Collar. $3.00-4.00
- ❏ ❏ #97 Pink Gift on Tail, Yellow Collar. $3.00-4.00
- ❏ ❏ #99 Red Santa Hat on Tail, Purple Collar. $3.00-4.00

Something on Head

- ❏ ❏ #36 Blue Bird on Head, Red/Orange Collar. $5.00-8.00
- ❏ ❏ #68 Green Frog on Head, Pink Collar. $5.00-8.00
- ❏ ❏ #70 Orange Butterfly on Head, Teal Collar. $5.00-8.00
- ❏ ❏ #82 Purple Bow on Head, Blue Collar. $5.00-6.00
- ❏ ❏ #85 Gray Wig on Head, Dark Pink/Fuchsia Collar. $5.00-6.00
- ❏ ❏ #89 Orange Maple Leaf on Head, Fuchsia Collar. $5.00-8.00
- ❏ ❏ #96 Green Cricket on Head, Purple Collar. $5.00-8.00

Comments: 101 Dalmatians were distributed in the USA stores.

Peter Pan Happy Meal, 1996

- ❏ ❏ Fra Pe9601 **Peter Pan Boat Viewer**. $3.00-5.00
- ❏ ❏ Fra Pe9602 **Captain Hook W Moveable Parts**. $3.00-5.00
- ❏ ❏ Fra Pe9603 **Alligator W Moving Tail**. $3.00-5.00
- ❏ ❏ Fra Pe9604 **Wendy W 6" Totem Pole**. $3.00-5.00
- ❏ ❏ Fra Pe9605 **Tinkerbell W Moving Wings**. $3.00-5.00

Comments: Distribution: France 1996.

Fra Pe9601-05

Fra Pe9626. French Display.

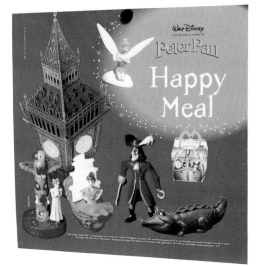

Fra Pe9664.
French Translite.

French Display.

Pocahontas Adventure Happy Meal, 1996

❏ ❏   Aus Po9601 **Pocahontas on Yellow Base W Play Scene**.
$4.00-5.00

❏ ❏   Aus Po9602 **Pocahontas on Blue Base W Play Scene**.
$4.00-5.00

❏ ❏   Aus Po9603 **Pocahontas on Pink Base W Play Scene**.
$4.00-5.00

❏ ❏   Aus Po9604 **Pocahontas on Gray Base W Play Scene**.
$4.00-5.00

Comments: Distribution: Australia - 1996. The four Pocahontas on Video Play Scene sets can be connected to build a fun and colorful play center.

Aus Po9601-04

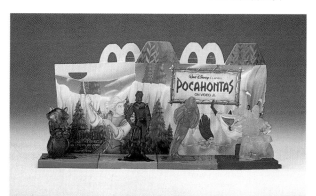

Australian Display.

1996

Smurfs 25 Years Germany Happy Meal, 1996

🖼	❑	Ger Sm9601 **Smurf Holding Present/Gift.**	$4.00-5.00
🖼	❑	Ger Sm9602 **Smurfette as a Cheerleader/Holding Pom Poms.**	$4.00-5.00
🖼	❑	Ger Sm9603 **Smurfette Marching.**	$4.00-5.00
🖼	❑	Ger Sm9604 **Smurf as a Clown.**	$4.00-5.00
🖼	❑	Ger Sm9605 **Smurf with Guitar.**	$4.00-5.00
❑	❑	Ger Sm9606 **Smurf Holding 25th Anniversary McDonald's Sign.**	$4.00-5.00
🖼	❑	Ger Sm9607 **Smurf with Milkshake.**	$4.00-5.00
🖼	❑	Ger Sm9608 **Smurf with Bread.**	$4.00-5.00
🖼	❑	Ger Sm9609 **Smurf with Big Mac.**	$4.00-5.00
🖼	❑	Ger Sm9610 **Smurf with Cake.**	$4.00-5.00

Comments: Distribution: Germany - September, 1996. All Smurfs have "McDonald's" logo somewhere on figurine.

Ger Sm9626

Ger Sm9601-05

Ger Sm9606-10

German Junior Tute Bag.

Ger Sm9664.
German Translite.

Counter Card, Holland.

Snoopy All Stars Happy Meal, 1996

❏	❏	Zea Sn9601 **#1 Snoopy Playing Tennis**.	$4.00-5.00
❏	❏	Zea Sn9602 **#2 Snoopy Playing Golf**.	$4.00-5.00
❏	❏	Zea Sn9603 **#3 Snoopy Playing Baseball**.	$4.00-5.00
❏	❏	Zea Sn9604 **#4 Snoopy Playing Soccer**.	$4.00-5.00

Comments: Distribution: New Zealand - 1996; Japan - 1997.

Zea Sn9601-04

Space Jam Happy Meal, 1996

❏	❏	USA Sp9601 **#1 Lola Bunny -** Lady Basketball Player W Org Basketball On Brown Base.	$2.00-2.50
❏	❏	USA Sp9602 **#2 Bugs Bunny -** Shooting Basket/Red Pole on Brown Base.	$2.00-2.50
❏	❏	USA Sp9603 **#3 Marvin the Martian -** Grn/Blk Martian Standing on Top of Org Ball on Brown Base.	$2.00-2.50
❏	❏	USA Sp9604 **#4 Daffy Duck -** Standing on Top of Orange Ball on Brown Base.	$2.00-2.50
❏	❏	USA Sp9605 **#5 Tasmanian Devil -** Shooting Ball into White Hoop on Brown Base.	$2.00-2.50
❏	❏	USA Sp9606 **#6 Monstar -** Green/Black Player W Arms Raised on Brown Base.	$2.00-2.50
❏	❏	USA Sp9607 **#7 Sylvester & Tweety -** Sylvester & Tweeety on Orange Ball on Brown Base.	$2.00-2.50
❏	❏	USA Sp9608 **#8 Nerdlucks -** Purple/Blue Totum Pole Style Player on Brown Base.	$2.00-2.50

Comments: National Distribution: USA - November 1-26, 1996; Germany (#1,2,3,4,5,7) - February, 1997.

USA Sp9601-04

USA Sp9605-08

1996

Ger Sp9626. German Display Dangler.

Ger Sp9664. German Translite.

Superman Metallic Happy Meal, 1996

❏ ❏	Aus Su9601 **Superman - Silver**.	$5.00-8.00
❏ ❏	Aus Su9602 **Superman - Gold with Cape**.	$5.00-8.00
❏ ❏	Aus Su9603 **Superman - Silver with Cape**.	$5.00-8.00
❏ ❏	Aus Su9604 **Lois**.	$5.00-8.00

Comments: Distribution: Australia - 1996.

Aus Su9601-04

Aus Su9626

68

Tonka Truck Happy Meal, 1996

❏	❏	Can Tk9405 **Wk 1 Loader** - Org W Blk Lift.	$2.00-3.00
❏	❏	Can Tk9406 **Wk 2 Crane** - Grn W Blk Hook.	$2.00-3.00
❏	❏	Can Tk9407 **Wk 3 Grader** - Yel W Yel Blade.	$2.00-3.00
❏	❏	Can Tk9408 **Wk 4 Bulldozer** - Yel W Blk Blade.	$2.00-3.00

Comments: National Distribution: Canada, Puerto Rico - December, 1994; New Zealand - 1996. Premiums are the same as: USA Cabbage Patch/Tonka Happy Meal 1994.

USA Tk9405-08

Fra To9607

Toy Story Happy Meal, 1996

❏	❏	Eur To9601 **Bo-Peep** - Spins on Base.	$4.00-5.00
❏	❏	Eur To9602 **Buzz Windup**.	$4.00-5.00
❏	❏	Eur To9603 **Hamm Bank** - Mini Piggy Bank.	$4.00-5.00
❏	❏	Eur To9604 **Rex dinosaur** - W Wobbly Head and Tail.	$4.00-5.00
❏	❏	Eur To9605 **Woody** - W Moveable Arms, Legs, and Head.	$4.00-5.00
❏	❏	Jpn To9606 **Toy Story Calendar**.	$5.00-8.00
❏	❏	Fra To9607 **Toy Story Pogs, each packet**.	$1.00-2.00

Comments: Distribution: Europe - March/April, 1996. Japan gave out four of the first five toys listed. Instead of Hamm a Calendar was distributed in Japan, November/December 1996. France distributed Toy Story Pogs.

Eur To9601-05

French Counter Card.

Fra To9626. Toy Story Display, France.

Toy Story Pog Board, France.

Fra To9664. French Translite.

Toy Story Music Cassettes/Musical Cassettes Happy Meal, 1996

☐ ☐ Sau To9601 **Cassette - You've Got A Friend in Me**.
$8.00-10.00
☐ ☐ Sau To9602 **Cassette - Strange Things**. $8.00-10.00
☐ ☐ Sau To9603 **Cassette - I Will Go Sailing No More**.
$8.00-10.00

Comments: Distribution: Saudia Arabia - Riad in 1996. Various musical cassettee have been produced and distributed around the world during the 1990s. Most cassettes were produced in a series of four to be distributed one per week for a month.

Saudia Arabia, English Version.

Saudia Arabia

Musical Cassettes advertising.

Saudia Arabia Pamphlet

1996

VR Troopers Happy Meal, 1996

❏ ❏   USA Vr9601 **#1 Visor -** Metal/Silver Plastic Eyewear.
$2.00-2.50

❏ ❏   USA Vr9602 **#2 Virtualizer -** Pendant W 2 Lenses.
$2.00-2.50

❏ ❏   USA Vr9603 **#3 Wrist Spinner -** Wristband W 2 Holographic
Discs/3p. $2.00-2.50

❏ ❏   USA Vr9604 **#4 Kaleidoscope -** Camera Shaped.
$2.00-2.50

Comments: National Distribution: USA - January 19-February 15, 1996.

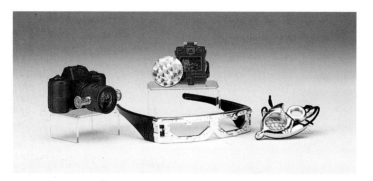

USA Vr9601-04

Walt Disney Home Video Masterpiece Collection I Happy Meal Act I, 1996

❏ ❏   USA Wa9601 **#1 Cinderella -** with Rooted Hair/Video Box.
$3.00-4.00

❏ ❏   USA Wa9602 **#2 Robin Hood W Gold Money Bag -** with
plush Tail/Gold Bag/**White Mouth**/Video Box. $3.00-4.00

❏ ❏   USA Wa9609 **#2 Robin Hood W Pea Green Money Bag
-** with plush Tail/Pea Green Bag/**Grey Mouth**/Video Box.
$3.00-5.00

❏ ❏   USA Wa9603 **#3 Pocahontas -** with Rooted Hair/Comb/
2p/Video Box. $3.00-4.00

❏ ❏   USA Wa9604 **#4 Aladdin -** with Moveable Arms, Head/
Waist/Video Box. $3.00-4.00

❏ ❏   USA Wa9605 **#5 Snow White -** with Rooted Hair/Video
Box. $3.00-4.00

❏ ❏   USA Wa9606 **#6 Merlin -** with Rooted Beard/Video Box.
$3.00-4.00

❏ ❏   USA Wa9607 **#7 Alice in Wonderland -** with Rooted Hair
And Comb/2p/Video Box. $3.00-4.00

❏ ❏   USA Wa9608 **#8 Scat Cat -** with Plush Stomach/Video Box.
$3.00-4.00

Comments: National Distribution: USA - April 19-May 16, 1996.

USA Wa9601-04

USA Wa9605-08

USA Wa9664. Translite.

USA Wa9602, 09

Willy Olympic Accessories Happy Meal, 1996

❏ ❏   Aus Wi9601 **Squeeze Ball**. $2.00-3.00
❏ ❏   Aus Wi9602 **Frisbee Flyer**. $2.00-3.00
❏ ❏   Aus Wi9603 **Baton - Red And Yellow**. $2.00-3.00

Comments: Distribution: Australia - July, 1996. Willy, the Australian Olympic Mascot, appeared on the toys.

Year of the Chicken Happy Meal, 1996

❏ ❏ Hon Ye9601 **Chicken -** Wearing Green Wrap. $4.00-5.00
❏ ❏ Hon Ye9602 **Chicken -** Wearing Blue Wrap. $4.00-5.00
❏ ❏ Hon Ye9603 **Chicken -** Wearing Light Blue Wrap.
$4.00-5.00
❏ ❏ Hon Ye9604 **Chicken -** Wearing Dark Blue Wrap.
$4.00-5.00

Comments: Distribution: Hong Kong - 1996.

Hon Ye9601-04

USA An9701-06

1997

• "My McDonald's" logo used

• 25th year of serving breakfast celebrated

• "Did somebody say McDonald's?" jingle

• McDonald's & Disney's "Flubber" is introduced

• Teenie Beanie Babies Happy Meal promotion is a sellout!

Animal Pals Insert Card.

Animal Pals Happy Meal, 1997

❏ ❏ USA An9701 **#1 Panda** - Blk/White Stuffed Bear.
$1.00-1.50
❏ ❏ USA An9702 **#2 Rhinoceros** - Grey Rhino W White Horn/ Stuffed. $1.00-1.50
❏ ❏ USA An9703 **#3 Yak** - Brn Yak W Wht Horns/Stuffed.
$1.00-1.50
❏ ❏ USA An9704 **#4 Moose -** Tan Moose W Wht Antlers/Stuffed.
$1.00-1.50
❏ ❏ USA An9705 **#5 Brown Bear** - Brown Small Baby Bear/ Stuffed. $1.00-1.50
❏ ❏ USA An9706 **#6 Gorilla -** Blk Gorilla W Wht Eyes/Nose/ Mouth/Stuffed. $1.00-1.50

❏ ❏ *** An9707 **Giraffe** - yellow/brown stuffed. $2.00-3.00

Comments: National Distribution: USA - August 1-August 24, 1997. Giraffe and parts of Amazing Wildlife from 1995 set were combined and distributed in countries around the world.

Jpn An9726. Japanese Display.

Barbie/Hot Wheels VIII Happy Meal, 1997

Premiums: Barbie Dolls

❑ ❑ USA Ba9701 **#1 Wedding Bride Rapunzel Barbie** - Wht Wedding Dress, Wht Hair, Wht Heart Doll Stand/2p.
$2.50-3.50

❑ ❑ USA Ba9702 **#2 Rapunzel Barbie** - Pnk Long Dress, White Hair, Grey Circle Doll Stand/2p. $2.50-3.50

❑ ❑ USA Ba9703 **#3 Angel Princess Barbie** - Wht Long Dress W Wht Petal Shaped Wings, Wht Hair, Pnk Doll Stand/2p.
$2.50-3.50

❑ ❑ USA Ba9704 **#4 Happy Holidays Barbie** - Red/White Long Dress, Long Brn Hair, Grn Wreath Doll Stand/2p.
$2.50-3.50

❑ ❑ USA Ba9705 **#5 Blossom Beauty Barbie** - African American, Pnk/Yel/Blue Floral Long Dress, Long Brn Hair, Yel Flower Petal Doll Stand/2p. $2.50-3.50

Premiums: Hot Wheels Vehicles

❑ ❑ USA Hw9706 **#6 Tow Truck** - Dark Blue Tow Truck W Yellow Lift & Bumper. $2.00-2.50

❑ ❑ USA Hw9707 **#7 Taxi Car** - Yellow Taxi W Black Spoiler.
$2.00-2.50

❑ ❑ USA Hw9708 **#8 Police Car** - Black/White Police Car W 1968 Gold Star on Hood. $2.00-2.50

❑ ❑ USA Hw9709 **#9 Ambulance Van** - White Ambulance W Red Flashing Lights/Red Cross Symbol on Side. $2.00-2.50

❑ ❑ USA Hw9710 **#10 Fire Truck** - Red Fire Truck W Grey Extension Hose on Top. $2.00-2.50

Comments: National Distribution: USA - October 24-November 27, 1997.

USA Ba9701-05

USA Hw9706-10

BeetleBorgs Metallix Happy Meal, 1997

❑ ❑ USA Be9701 **#1 Beetle Bonder** - Gry/Blk W Blk Claws, Wings Flip Out. $2.00-2.50

❑ ❑ USA Be9702 **#2 Chromium Gold Beetleborg Covert Compact** - Gold/Blue Compact W Blk Cord/Opens/W Adv Booklet. $2.00-2.50

❑ ❑ USA Be9703 **#3 Hunter Claw** - Gry/Blk Beetle Shaped W Orange Claws/Top Flips Up. $2.00-2.50

❑ ❑ USA Be9704 **#4 Platinum Purple Beetleborg Covert Compact** - Purp/Red Compact W Clasp/Opens.
$2.00-2.50

❑ ❑ USA Be9705 **#5 Stinger Drill** - Blue/Blk Drill/Cone Shape W Crank. $2.00-2.50

❑ ❑ USA Be9706 **#6 Titanium Silver Beetleborg Covert Compact Wristband -** Bright Silver Compact W Silver Claws/Opens. $2.00-2.50

Comments: National Distribution: USA - August 15-September 4, 1997.

USA Be9701-03

USA Be9704-06

Cartoon Carnival Happy Meal, 1997

❑ ❑	*** Cc9701	**Popeye on Vehicle.**	$7.00-10.00
❑ ❑	*** Cc9702	**Woody Woodpecker in Vehicle.**	
			$7.00-10.00
❑ ❑	*** Cc9703	**Tom and Jerry in Vehicle.**	$7.00-10.00
❑ ❑	*** Cc9704	**Mighty Mouse in Vehicle.**	$7.00-10.00

Comments: Distribution is uncertain, may be a prototype distributed around the world.

Cc9701-04

Cartoon Network Happy Meal, 1997

❑ ❑	Aus Ca9701	**Fred Flintstone Bowl.**	$3.00-4.00
❑ ❑	Aus Ca9702	**Yogi Bear Bowl.**	$2.00-4.00
❑ ❑	Aus Ca9703	**Magilla Gorilla Placemat.**	$2.00-4.00
❑ ❑	Aus Ca9704	**Flitstone Sipper Cup.**	$2.00-4.00

Comments: Distribution: Australia - November, 1997.

Aus Ca9701-04

Disney on Parade Happy Meal, 1997

❑ ❑	Zea Mi9404	**Donald in Mexico.**	$2.00-2.50
❑ ❑	Zea Mi9405	**Goofy in Norway.**	$2.00-2.50
❑ ❑	Zea Mi9406	**Mickey in USA.**	$2.00-2.50
❑ ❑	Zea Mi9407	**Minnie in Japan.**	$2.00-2.50

Comments: Distribution: New Zealand - 1997. The USA distributed four additional characters: Chip, Daisy, Dale, and Pluto with Mickey & Friends Happy Meal in 1994 (Mi9401-04).

USA Mi9401-04

USA Mi9405-08

Disneyland Adventure Happy Meal, 1997

❏ ❏ Jpn Di9701 **Donald** - In Yellow House. $5.00-6.00
❏ ❏ Jpn Di9702 **Minnie's Cottage** - Pink. $5.00-6.00
❏ ❏ Jpn Di9703 **Daisey** - Pink Cottage. $5.00-6.00
❏ ❏ Jpn Di9704 **Mickey's Adventure House** - Blue and White.
$5.00-6.00

Comments: Distribution: Japan - May, 1997.

Jpn Di9701-04

Disneyland Water Adventure Happy Meal, 1997

❏ ❏ Kor Di9701 **Mickey** - In Blue Water Craft. $5.00-6.00
❏ ❏ Kor Di9702 **Goofy** - In Blue Jalopy. $5.00-6.00
❏ ❏ Kor Di9703 **Chip & Dale** - In Red, Blue Windup.
$5.00-6.00
❏ ❏ Kor Di9704 **Minnie** - In Brown Boat Holding Paddle.
$5.00-6.00
❏ ❏ Kor Di9705 **Donald** - Riding Dumbo. $5.00-6.00
❏ ❏ Kor Di9706 **Pluto** - In Log Boat. $5.00-6.00

Comments: Distribution: Korea - 1997.

Kor Di9701-06

Fisher-Price Toddler Toys for U-3, 1997

Group G
❏ ❏ USA Fi9701 **Ronald in a Green/Purple Boat.**
$3.00-4.00
❏ ❏ USA Fi9702 **Hamburglar in a Burgermobile Walker W Blue Wheels.** $3.00-4.00
❏ ❏ USA Fi9703 **Boy Pilot W Orange Scarf in White/Red Airplane.** $3.00-4.00
❏ ❏ USA Fi9704 **Farmer Boy W Yellow Hat in Red Tractor W Green Wheels.** $3.00-4.00

Group H
❏ ❏ USA Fi9705 **Rolling Ball** - Green W White/Orange Rolling Face. $3.00-4.00
❏ ❏ USA Fi9706 **Ring Toss** - White Base W Yellow & Blue Rings/3p. $3.00-4.00
❏ ❏ USA Fi9707 **Phone** - White/Blue Cell Phone With Purp/Green/Orange/Yellow Push Knobs. $3.00-4.00
❏ ❏ USA Fi9708 **Radio** - Red/Yellow Square Radio W Blue Dial. $3.00-4.00

Group I
❏ ❏ USA Fi9709 **Train Engineer** - Blue in Blue Train Engine.
$3.00-4.00
❏ ❏ USA Fi9710 **Grimace in Store Doorway W Revolving Door.** $3.00-4.00
❏ ❏ USA Fi9711 **Hamburglar in Flat Hamburger Magnet.**
$3.00-4.00
❏ ❏ USA Fi9712 **Lawn Mower Push Toy W Character Pictured/Musical Instruments Rolling On Red Wheels.**
$3.00-4.00

Group J
❏ ❏ USA Fi9713 **Car W Dog** - Spotted White/Black on Roller Ball. $3.00-4.00
❏ ❏ USA Fi9714 **Dog** - Blue W Blue Collar/Yellow Bell.
$3.00-4.00
❏ ❏ USA Fi9715 **Elephant Rolling Car** - Grey W White Baby Elephant Car. $3.00-4.00
❏ ❏ USA Fi9716 **Camera** - Blue/Red/White W Red/Yellow Dial. $3.00-4.00

Group K
❏ ❏ USA Fi9715 **Silo** - Red/Yellow/Blue. $3.00-4.00
❏ ❏ USA Fi9716 **Taxi** - Yellow W "Taxi" on Door.
$3.00-4.00
❏ ❏ USA Fi9717 **Dog in White Car W Blue Rolling Figure.**
$3.00-4.00
❏ ❏ USA Fi9718 **Cow.** $3.00-4.00

Group L
❏ ❏ USA Fi9721 **Chicken McNuggets in Box.** $3.00-4.00
❏ ❏ USA Fi9722 **Soft Drink Cup W Straw.** $3.00-4.00
❏ ❏ USA Fi9723 **Apple Pie in Red Box.** $3.00-4.00
❏ ❏ USA Fi9724 **French Fries in White Package.** $3.00-4.00

❏ ❏ USA Fi9727 **Color Card** - picture of all 24 Fisher Price toys. $3.00-5.00

Comments: Distribution: USA - September 1997-August 1998. Portions of the above Fisher-Price U-3 toys were distributed in numerous countries around the world, along with the 1996 set of 24 different Fisher-Price toys.

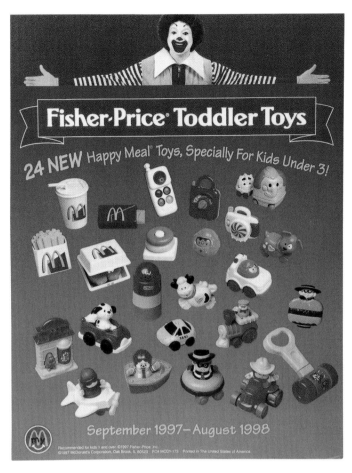

USA Fp9701-24

Ger Pa9764. German Translite.

Five Years Disneyland Paris Happy Meal, 1997

☐	☐	Ger Pa9701 **Goofy on Sectional Piece**.	$3.00-5.00
☐	☐	Ger Pa9702 **Donald on Sectional Piece**.	$3.00-5.00
☑	☐	Ger Pa9703 **Mickey on Sectional Piece**.	$3.00-5.00
☐	☐	Ger Pa9704 **Minnie on Sectional Piece**.	$3.00-5.00
☐	☐	Ger Pa9705 **Pluto on Sectional Piece**.	$3.00-5.00

Comments: Distribution: Germany - May, 1997.

Ger Pa9701-05

Ger Pa9726. German Display Dangler.

Flying Dragons/Les Dragons Volants/Zauberdrachen Happy Meal, 1997

- ☑ ❏   Ger Di9701 **Dinosaur Skeleton -** blue on orange stand/3p. $2.00-3.00
- ❏ ❏   Ger Di9702 **Dinosaur Skeleton -** green on purple stand/3p. $2.00-3.00
- ❏ ❏   Ger Di9703 **Dinosaur Skeleton -** purple on green stand/3p. $2.00-3.00
- ❏ ❏   Ger Di9704 **Dinosaur Skeleton -** orange on pink stand/3p. $2.00-3.00

- ❏ ❏   Uk Di9705 **Sleeping Dragon -** blue on green stand/3p. $2.00-3.00
- ❏ ❏   Uk Di9706 **Water Dragon -** green on blue stand/3p. $2.00-3.00
- ❏ ❏   Uk Di9707 **Mountain Peak Dragon -** yellow on pink stand/3p. $2.00-3.00
- ❏ ❏   Uk Di9708 **Tower -** blue on orange stand/3p. $2.00-3.00

Comments: Distribution: Germany - 1997; UK - September. 1998. Base color changed around the world.

Ger Di9701-04. Insert Card.

Uk Di9705-08

Flying Dragons Insert Card.

Fra Di9726. French Display.

Fra Di9764. French Translite.

Hercules Happy Meal, 1997

New Zealand Premiums:

- ❏ ❏ USA He9702 **#2 Rock Titan & Zeus -** Org/Purple Figure/ Brown Case/3p. $2.00-2.50
- ❏ ❏ USA He9703 **#3 Hydra & Hercules -** Brown/Tan Figure/ Gold Shield/Purple Case/3p. $2.00-2.50
- ❏ ❏ USA He9707 **#7 Pegasus & Megara -** Female Figure/Purp Dress/Long Hair/Wht/Blu Horse Case/3p. $2.00-2.50
- ❏ ❏ USA He9709 **#9 Nessus & Phil -** Tan/Org Figure/Dark Purp Case/3p. $2.00-2.50

USA Additional Premiums:

- ❏ ❏ USA He9701 **#1 Wind Titan & Hermes -** Blue Figure/Grey Case/3p . $2.00-2.50
- ❏ ❏ USA He9704 **#4 Lava Titan & Baby Pegasus -** White/ Blue Baby Horse/Red Case. $2.00-2.50
- ❏ ❏ USA He9705 **#5 Cyclops & Pain -** Purple Figure/Lg White Teeth/Cream Case/3p. $2.00-2.50
- ❏ ❏ USA He9706 **#6 Fates & Panic -** Turq Figure/Wht Eyeball/ Grey Case/3p. $2.00-2.50
- ❏ ❏ USA He9708 **#8 Ice Titan & Calliope -** Female Figure/ Purp Dress/Short Brn Hair/Clear Case/3p . $2.00-2.50
- ❏ ❏ USA He9710 **#10 Cerberus & Hades -** Blue/Grey Figure/ Blk Red Eye/3 Headed Case/3p. $2.00-2.50

Comments: Distribution: New Zealand - 1997, four premiums. The USA distributed an additional six premiums, for a total of ten Hercules toys in the USA.

USA He9701-03

USA He9704-06

USA He9707-10

Hunchback of Notre Dame Happy Meal, 1997

- ❏ ❏ USA Hu9701 **#1 Esmeralda Amulet -** Necklace W Tan Amulet Hanging on Purp Cord. $1.00-1.50
- ❏ ❏ USA Hu9702 **#2 Scepter -** Org/Red Scepter W Push But-ton/Quasimodo Appears Inside. $1.00-1.50
- ❏ ❏ USA Hu9703 **#3 Clopin Mask -** Purp/Pnk Mask on Purple Pole. $1.00-1.50
- ❏ ❏ USA Hu9704 **#4 Hugo Horn -** Hugo/Grey Blowing Purple/ Yel Whistle/Horn. $1.00-1.50

USA Hu9701-04

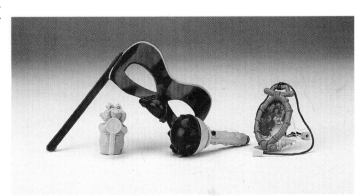

❑ ❑ USA Hu9705 **#5 Juggling Balls** - 3 Grey Rubber Balls/Faces on Balls/3p. $1.00-1.50

❑ ❑ USA Hu9706 **#6 Drum** - Jester W Purp Hat on Top of Purp/Pnk Drum W Base. $1.00-1.50

❑ ❑ USA Hu9707 **#7 Quasimodo Bird Catcher** - Quasimodo in Green Shirt W Blue Bird in Brn Dish. $1.00-1.50

❑ ❑ USA Hu9708 **#8 Tambourine** - Blue/Orange W 2 Green Ribbons. $1.00-1.50

Comments: National Distribution: USA - March 7-April 3, 1997.

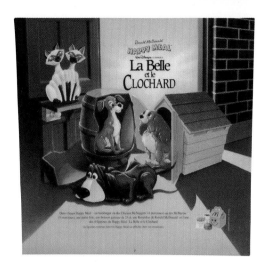

Fra La9764. French Translite.

USA Hu9705-08

Jungle Book II/Halloween '97 Happy Meal, 1997

Premiums: Figurines with Nerds Candy Packet (MIP)

❑ ❑ USA Ju9701 **#1 Baloo** - Bear Holding Yellow Bunch Of Bananas W Grape Nerds Candy. $2.00-2.50

❑ ❑ USA Ju9702 **#2 Junior** - Elephant W Cherry Nerds candy. $2.00-2.50

❑ ❑ USA Ju9703 **#3 Bagheera** - Cheta W Rainbow Nerds candy. $2.00-2.50

❑ ❑ USA Ju9704 **#4 King Louie** - Monkey W Rainbow Nerds candy. $2.00-2.50

❑ ❑ USA Ju9705 **#5 Kaa** - Snake on Palm Tree W Watermelon Nerds candy. $2.50-3.00

❑ ❑ USA Ju9706 **#6 Mowgli** - Boy and Coconuts W Strawberry Nerds candy. $2.00-2.50

Comments: National Distribution: USA - October 3-23, 1997. Kaa was recalled towards the end of the promotion.

Lady and The Tramp Happy Meal, 1997

❑ ❑ Zea La9701 **Lady**. $3.00-4.00
❑ ❑ Zea La9702 **Tramp**. $3.00-4.00
❑ ❑ Zea La9703 **Si & Am**. $3.00-4.00
❑ ❑ Zea La9704 **Trusty**. $3.00-4.00

Comments: Distribution: New Zealand, France, and Germany - 1997.

Ger La9726. Display Section.

Fra La9701-04

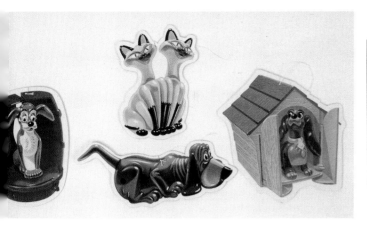

Ger La9701-04

USA Li9705-08

Little Mermaid II Happy Meal, 1997

❑ ❑ USA Li9701 **Ursula** - Black/Purple Ursula Blowup Floater W Grey Ring/2p. $3.00-4.00

❑ ❑ USA Li9702 **Flounder** - Yellow/Blue Fish Floater. $3.00-4.00

❑ ❑ USA Li9703 **Scuttle** - White Bird W Orange Feet/Windup Floater. $3.00-4.00

❑ ❑ USA Li9704 **Ariel** - Mermaid W Red Hair Holding Seahorse/ Floater. $3.00-4.00

❑ ❑ USA Li9705 **Max** - Grey Sea Animal W 2 Arms/Floater. $3.00-4.00

❑ ❑ USA Li9706 **Glut** - Grey Shark/Floater. $3.00-4.00

❑ ❑ USA Li9707 **Eric** - Prince Eric in Boat/Floater. $3.00-4.00

❑ ❑ USA Li9708 **Sebastian** - Red Crab/Floater. $3.00-4.00

❑ ❑ USA Li9709 **Gold Ursula** - Black/Purple Ursula Blowup Floater W Gold Ring/2p. $4.00-5.00

❑ ❑ USA Li9710 **Gold Flounder** - Gold Fish Floater. $4.00-5.00

❑ ❑ USA Li9711 **Gold Scuttle** - Gold Windup Floater. $4.00-5.00

❑ ❑ USA Li9712 **Gold Ariel** - Gold Mermaid Floater. $4.00-5.00

❑ ❑ USA Li9713 **Gold Max** - Gold Sea Animal W 2 Arms/Floater. $4.00-5.00

❑ ❑ USA Li9714 **Gold Glut** - Gold Shark/Floater. $4.00-5.00

❑ ❑ USA Li9715 **Gold Eric** - Gold Prince Eric in Boat/Floater. $4.00-5.00

❑ ❑ USA Li9716 **Gold Sebastian** - Gold Crab/Floater. $4.00-5.00

Comments: National Distribution: USA: November 28-December 25, 1997. One of every ten toys distributed was reported to be one of the "Gold Set." Gold Sets were sold by McDonald's through a coupon in the stores, to be delivered at a later date.

McDonaldland Space 3000 Happy Meal, 1997

❑ ❑ Zea Sc9701 **Super Sparker**. $3.00-4.00
❑ ❑ Zea Sc9702 **Triangle Trekker**. $3.00-4.00
❑ ❑ Zea Sc9703 **Saucer Spinner**. $3.00-4.00
❑ ❑ Zea Sc9704 **Creeper Crawler**. $3.00-4.00

Comments: Distribution: New Zealand - 1997.

Zea Sc9701-04

McDonald's Jubilee Happy Meal, 1997

❑ ❑ *** Ju9701 **Hatspin Grimace** - Purple W Twirler on Hat. $5.00-8.00

❑ ❑ *** Ju9702 **Hightrailin Hamburglar** - W Flipout Wheels. $5.00-8.00

❑ ❑ *** Ju9703 **Gyro CosMc** - W Zip Strip. $5.00-8.00

❑ ❑ *** Ju9704 **Flap Flying Birdie** - W Pull String. $5.00-8.00

Comments: Distribution: Unconfirmed prototype toys and/or Southeast Asia.

USA Li9701-04

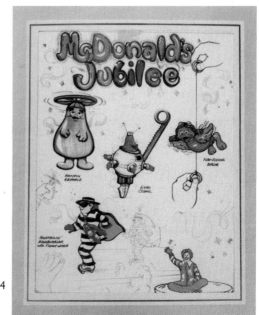

***Ju9701-04

McFun Park/McDonaldland Figures & Playground Happy Meal, 1997

		Zea Pa9701 **Birdie's Seesaw**.	$6.00-10.00
❏	❏	Zea Pa9701 **Birdie's Seesaw**.	$6.00-10.00
❏	❏	Zea Pa9702 **Grimace's Swingset**.	$6.00-10.00
❏	❏	Zea Pa9703 **Hamburglar's Roundabout**.	$6.00-10.00
❏	❏	Zea Pa9704 **Ronald's Slippery Slide**.	$6.00-10.00

Comments: Distribution: New Zealand - 1998. This set was originally released in Australia in 1997 as McDonaldland Figures & Playground. The Ronald figure in particular has been used during clean-up in many countries around the world.

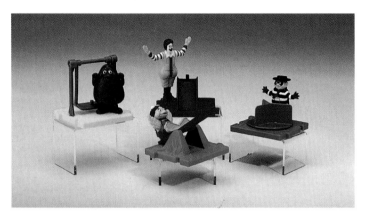

Zea Pa9701-04

Mighty Ducks the Animated Series Happy Meal, 1997

❏	❏	USA Mi9701 **#1 Wildwing -** Black Hockey Puck W White Beak Duck.	$1.00-1.50
❏	❏	USA Mi9702 **#2 Nosedive -** Purple Hockey Puck W Orange Beak Duck.	$1.00-1.50
❏	❏	USA Mi9703 **#3 Mallory -** Yellow Hockey Puck W Female Duck/Org Beak.	$1.00-1.50
❏	❏	USA Mi9704 **#4 Duke L'orange -** Blue Hockey Puck W Orange Beak.	$1.00-1.50

Comments: National Distribution: USA - February 7-March 6, 1997.

USA Mi9701-04

Oliver & Company Happy Meal, 1997

	❏	Uk Ol9701 **Tito the Dog**.	$4.00-5.00
	❏	Uk Ol9702 **Roger on the Motorbike**.	$4.00-5.00
	❏	Uk Ol9703 **Jenny and Oliver**.	$4.00-5.00
	❏	Uk Ol9704 **Fagin on the car**.	$4.00-5.00

Comments: Distribution: UK - September, 1999.

Uk Ol9701-04. Oliver & Company Insert Card.

Uk Ol9701-04. Oliver & Company Trayliner.

101 One Hundred and One Dalmatians Flipcars Happy Meal III, 1997

Premiums:

❏ ❏ USA On9701 **Toy #1 Grey Dalmatian** - in Orange/Blue Flip Car W pink wheels. $2.00-2.50

❏ ❏ USA On9702 **Toy #2 White Dalmatian** - W 2 Blk Ears In Yellow/Purp Flip Car W Blk/Org Wheels. $2.00-2.50

❏ ❏ USA On9703 **Toy #3 White Dalmatian** - W 2 Blk Ears/Red Collar In Grey/Org Flip Car W Blk/Grn Wheels. $2.00-2.50

❏ ❏ USA On9704 **Toy #4 White Dalmatian** - W Spotted Ears/ Blue Collar in Light Purple/Grey Flip Car W Org/Blue Wheels. $2.00-2.50

❏ ❏ USA On9705 **Toy #5 Green Dalmatian** - in Blu/Purp/Grn Flip Car W Yellow/Org Wheels. $2.00-2.50

❏ ❏ USA On9706 **Toy #6 Pink Dalmatian** - in Org/Green Flip Car W White/Blue Wheels. $2.00-2.50

❏ ❏ USA On9707 **Toy #7 White Dalmatian** - W Blue Collar in Blue/Yellow Flip Car W Grn/Purp Wheels. $2.00-2.50

❏ ❏ USA On9708 **Toy #8 Grey Dalmatian** - in Yellow/Purple Flip Car W Red/Tan Wheels. $2.00-2.50

Comments: National Distribution: USA - December 26, 1997-January 22, 1998.

USA On9701-04

USA On9705-08

Play-Doh Happy Meal, 1997

❏ ❏ *** Mo9701 **Play-Doh** - Red W Hamburglar Mold. $2.00-4.00

❏ ❏ *** Mo9702 **Play-Doh** - Blue W Birdie Mold. $2.00-4.00
❏ ❏ *** Mo9703 **Play-Doh** - Green W Grimace Mold. $2.00-4.00

❏ ❏ *** Mo9704 **Play-Doh** - Yellow W Ronald Mold. $2.00-4.00

❏ ❏ Hol Mo9705 **Play-Doh** - Red W Hamburglar Mold. $2.00-4.00

❏ ❏ Hol Mo9706 **Play-Doh** - Orange W Birdie Mold. $2.00-4.00

❏ ❏ Hol Mo9707 **Play-Doh** - Green W Grimace Mold. $2.00-4.00

❏ ❏ Hol Mo9708 **Play-Doh** - Yellow W Ronald Mold. $2.00-4.00

❏ ❏ *** Mo9709 **Play-Doh** - Red W Dinosaur Mold. $2.00-4.00

❏ ❏ *** Mo9710 **Play-Doh** - Blue W Dinosaur Mold. $2.00-4.00

❏ ❏ *** Mo9711 **Play-Doh** - Green W Dinosaur Mold. $2.00-4.00

❏ ❏ *** Mo9712 **Play-Doh** - Yellow W Dinosaur Mold. $2.00-4.00

❏ ❏ *** Mo9713 **Play-Doh** - Blue & Pink 2 Small Cans W Mold. $2.00-4.00

Comments: Distribution: Latin America, Far East - 1997. Each canister came with a mold. Various countries around the world distributed Play-Doh canisters with various designed molds.

***Mo9701-04

Hol Mo9705-08   ***Mo9709-12

***Mo9713-16

1997

Isr Mo9517-20

Ger Po9764. German Translite.

Pocket Macs/McPockets Happy Meal, 1997

📷	☐	Ger Po9701 **Hamburglar -** in Hamburger.	$4.00-5.00
📷	☐	Ger Po9702 **Grimace -** in Milkshake.	$4.00-5.00
📷	☐	Ger Po9703 **Ronald -** in Ice Cream Sundae.	$4.00-5.00
📷	☐	Ger Po9704 **Birdie -** in French Fries.	$4.00-5.00

Comments: Distribution: Germany - April, 1997.

Fra Po9726. French Display for McPockets.

Ger Po9701-04

Fra Po9764. French
Translite.

Safari/Jungle Safari Happy Meal, 1997

❑	❑	Zea Sa9701	**Ronald in Jeep.**	3.00-4.00
❑	❑	Zea Sa9702	**Birdie on Raft.**	3.00-4.00
❑	❑	Zea Sa9703	**Hamburglar in Tree Hut.**	3.00-4.00
❑	❑	Zea Sa9704	**Grimace on Elephant.**	3.00-4.00

Comments: Distribution: New Zealand - 1997; South Africa - November/December, 1995. In South Africa this was the first Happy Meal called "Kid's Favorite Meal Around the World."

Zea Sa9701-04

Santa & Friends Decorations Happy Meal, 1997

❑	❑	Aus Sa9701 **Santa with Pack/Black Shoes** - blowup.	
			$3.00-4.00
❑	❑	Aus Sa9702 **Candy Cane/Dr. Seuss Style** - blowup.	
			$3.00-4.00
❑	❑	Aus Sa9703 **Dog on Sled** - blowup.	$3.00-4.00
❑	❑	Aus Sa9704 **Sleigh** - blowup.	$3.00-4.00
❑	❑	Aus Sa9705 **Gift** - blowup.	$3.00-4.00
❑	❑	Aus Sa9706 **Christmas Tree** - blowup.	$3.00-4.00
❑	❑	Aus Sa9707 **Snowman** - blowup.	$3.00-4.00
❑	❑	Aus Sa9708 **Santa with Gift** - blowup.	$3.00-4.00

Comments: Distribution: Australia - December, 1997.

Aus Sa9702, 04, 06, 07

Aus Sa9708, 03, 01, 05

Seusmobiles/Dr. Seuss Happy Meal, 1997

❑	❑	*** Se9701	**Horton the Elephant in Auto.**	
				$10.00-15.00
❑	❑	*** Se9702	**The Cat in the Hat in Auto.**	$10.00-15.00
❑	❑	*** Se9703	**One Fish Two Fish in Auto.**	$10.00-15.00
❑	❑	*** Se9704	**Sam I Am in Auto.**	$10.00-15.00
❑	❑	*** Se9705	**The Grinch in Auto.**	$10.00-15.00
❑	❑	*** Se9706	**Fox in Socks in Auto.**	$10.00-15.00

Comments: Distribution: Uncertain. May be prototype Happy Meal toys.

***Se9701-06

Seus Mix N Match/Dr. Seuss Happy Meal, 1997

❏	❏	*** Sm9701	**The Cat in the Hat**.	$10.00-15.00
❏	❏	*** Sm9702	**One Fish Two Fish**.	$10.00-15.00
❏	❏	*** Sm9703	**Fox the Socks**.	$10.00-15.00
❏	❏	*** Sm9704	**Yertle the Turtle**.	$10.00-15.00
❏	❏	*** Sm9705	**The Grinch**.	$10.00-15.00
❏	❏	*** Sm9706	**Thidwick the Big-Hearted Moose**.	
				$10.00-15.00
❏	❏	*** Sm9707	**Horton the Elephant**.	$10.00-15.00
❏	❏	*** Sm9708	**Sam I Am**.	$10.00-15.00

Comments: Distribution: Uncertain. May be prototype Happy Meal toys.

***Su9701-08

Sippers Happy Meal, 1997

❏	❏	Ger Si9701	**Sipper Cup - Red**.	$3.00-5.00
❏	❏	Ger Si9702	**Sipper Cup - Yellow**.	$3.00-5.00
❏	❏	Ger Si9703	**Sipper Cup - Black**.	$3.00-5.00
❏	❏	Ger Si9704	**Sipper Cup - Blue**.	$3.00-5.00

Comments: Distribution: Germany, Ireland - 1997.

Ger Si9701-04

Sky Dancers and Micro Machines Happy Meal, 1997

Premiums For Girls: Dancing Dolls

❏ ❏ USA Sk9701 **#1 Swan Shimmer -** Elegant Ballerina, Nd, Brn Face, Org Wings/Body W Lt Blue Base/2p. $2.00-2.50

❏ ❏ USA Sk9702 **#2 Rosemerry -** Rosemarie, Nd, Red Hair, Wht Face, Lime Green Wings W Dk Purp Base/2p.
$2.00-2.50

❏ ❏ USA Sk9703 **#3 Flutterfly -** Luciole, Nd, Pnk Hair, Wht Face, Lt Yel Wings W Turq Base/2p. $2.00-2.50

❏ ❏ USA Sk9704 **#4 Princess Pegus -** Princess Pivoine, Nd, Yel Hair, Wht Face, Pnk Wings W Lt Purp Base/2p. $2.00-2.50

Premiums For Boys: Vehicles

❏ ❏ USA Sk9705 **#5 Evac Copter -** Gold Helicopter Body/Blk Wings/Red Base. $2.00-2.50

❏ ❏ USA Sk9706 **#6 Polar Explorer Vehicle -** Lt Purp/Silver Body W Red/Blue Front Bumper/Antennae on Roof.
$2.00-2.50

❏ ❏ USA Sk9707 **#7 Deep Sea Hunter Sea Crane -** Grey Crane/Metallic Org Body/Blu Base. $2.00-2.50

❏ ❏ USA Sk9708 **#8 Ocean Flyer Airplane Sea Plane -** Silver Body/Blk Prop/Red Base. $2.00-2.50

Comments: National Distribution: USA - May 30-June 19, 1997.

USA Sk9701-04

USA Sk9705-08

Sleeping Beauty Happy Meal, 1997

Premiums:

❏ ❏ USA SI9701 **#1 Sleeping Beauty Pencil Cap & Eraser** - Sleeping Beauty W Long Blue Dress W Yel Spinning Wheel Eraser/2p. $2.00-3.00

❏ ❏ USA SI9702 **#2 Maleficent Ruler & Stencil** - Black Cape Maleficent W Green Ruler/Stencil/2p. $2.00-3.00

❏ ❏ USA SI9703 **#3 Prince Philip** - Holding Blue Shield W Grey Sword/3p. $2.00-3.00

❏ ❏ USA SI9704 **#4 Flora Paper-Punch** - Orange/Grey Rolly Polly Flora W Inset Paper Punch. $2.00-3.00

❏ ❏ USA SI9705 **#5 Dragon Ink Pen** - Black/Grey/Purple Dragon W Inset Ink Pen. $2.00-3.00

❏ ❏ USA SI9706 **#6 Raven Book Clip** - Black Raven W Wings As Clip. $2.00-3.00

Comments: National Distribution: USA - September 12-October 2, 1997.

USA SI9701-06

Splash Time Happy Meal, 1997

❏ ❏ *** Sp9701 **Ronald Takes a Dive.** $10.00-15.00

❏ ❏ *** Sp9702 **Birdie Paddles Her Way Along.** $10.00-15.00

❏ ❏ *** Sp9703 **Fry Kids on Dolphin.** $10.00-15.00

❏ ❏ *** Sp9704 **Grimace Skims Over Surface Like Hover Craft.** $10.00-15.00

Comments: Distribution: Uncertain. May be prototype toys.

***Sp9701-04

Sports Frisbees Happy Meal, 1997

❏ ❏ Zea Fr9701 **Frisbee - Ronald Running.** $1.00-2.00

❏ ❏ Zea Fr9702 **Frisbee - Grimace Weight Lifting.** $1.00-2.00

❏ ❏ Zea Fr9703 **Frisbee - Hamburglar.** $1.00-2.00

❏ ❏ Zea Fr9704 **Frisbee - Birdie.** $1.00-2.00

Comments: Distribution: New Zealand - 1997. The four frisbees feature the four McDonaldland characters in sporting modes.

Super Shifters Happy Meal, 1997

❏ ❏ Zea Su9701 **Ronald on Green Vehicle.** $4.00-5.00

❏ ❏ Zea Su9702 **Birdie on Pink Vehicle.** $4.00-5.00

❏ ❏ Zea Su9703 **Grimace on Red Vehicle.** $4.00-5.00

❏ ❏ Zea Su9704 **Hamburglar on Yellow Vehicle.** $4.00-5.00

Comments: Distribution: New Zealand - 1997; Japan - April, 1997.

Tangle Twist-A-Toid Happy Meal, 1997

❏ ❏ USA Ta9701 **#1 - Purple/Pink 3-Legged Creature** W 2 Long Pink/2 Short Pink Twists/5p. $1.00-2.00

❏ ❏ USA Ta9702 **#2 - Lime Green Creature** W 2 Long Orange/2 Short Orange Twists/5p. $1.00-2.00

❏ ❏ USA Ta9703 **#3 - Red Body/Gold Eyed Headed Standing Creature** W 2 Long Purple/2 Short Purple Twists/5p. $1.00-2.00

❏ ❏ USA Ta9704 **#4 - Gold Body Creature W 3 legs** W 2 Lime Grn/1 Blue/1 Orange Twist/5p. $1.00-2.00

USA Ta9701-04

❏ ❏ USA Ta9705 **#5 - Purple Body Creature W Lime Green Eyes** W 1 Long Grn/1 Short Grn/1 Gold Short/1 Pnk Short Twist/5p. $1.00-2.00

❏ ❏ USA Ta9706 **#6 - Orange Body/Purple Legged Creature** W 1 Long Gold/1 Short Gold/1 Long Grn/1 Short Purple Twist/5p. $1.00-2.00

❏ ❏ USA Ta9707 **#7 - Green Body Creature** Standing W 1 Long Purple/1 Short Purple/1 Short Grn/1 Short Purple/5p. $1.00-2.00

❏ ❏ USA Ta9708 **#8 - Blue Blow Fish Creature** W 1 Long Org/1 Short Org/1 Long Blue/1 Short Purple Twist/5p. $1.00-2.00

Comments: National Distribution: USA - January 10-February 6. 1997.

USA Ta9705-08

Teenie Beanie Babies I Happy Meal, 1997

Premiums:

❏ ❏ USA Ty9701 **#1 Patti Platypus -** Purple Stuffed Platypus W Gold Beak/Feet. $10.00-15.00

❏ ❏ USA Ty9702 **#2 Chops Lamb -** White Stuffed Lamb W Black Face. $10.00-15.00

❏ ❏ USA Ty9703 **#3 Goldie Goldfish -** Gold Goldfish W Red Tale Stripe. $10.00-15.00

USA Ty9701-03

❏ ❏ USA Ty9704 **#4 Seamore Seal -** White Seal W Black Eyes. $5.00-10.00

❏ ❏ USA Ty9705 **#5 Quacks Duck -** Yellow Duck W Gold Beak/Feet. $5.00-10.00+

❏ ❏ USA Ty9706 **#6 Pinky Flamingo -** Pink Flamingo W Gold Beak. $5.00-10.00+

❏ ❏ USA Ty9707 **#7 Chocolate Moose -** Brown Moose with Gold Antlers. $5.00-10.00+

❏ ❏ USA Ty9708 **#8 Speedy Turtle -** Green Turtle with Brown Shell. $5.00-10.00+

❏ ❏ USA Ty9709 **#9 Snort Bull -** Red Bull W White Hoofs. $5.00-10.00+

❏ ❏ USA Ty9710 **#10 Lizzy Lizard -** Blue Lizard with Black Spots/Gold Stomach. $5.00-10.00+

Comments: National Distribution: USA - April 11-May 15, 1997.

USA Ty9704-06

USA Ty9707, 08, 10

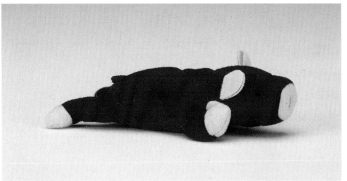

USA Ty9709

Walt Disney 25th Anniversary Celebration Happy Meal, 1997

❏ ❏   Jpn An9701 **Donald Duck on Parade.** $4.00-6.00
❏ ❏   Jpn An9702 **Sorcerer Mickey with Castle Cake.**
                    $4.00-6.00
❏ ❏   Jpn An9703 **Minnie Mouse (Enchanted) in Coach.**
                    $4.00-6.00
❏ ❏   Jpn An9704 **Goofy as Celebration Surprise.** $4.00-6.00

Comments: Distribution: Japan - July 1997

Walt Disney Home Video Masterpiece Collection Act II Happy Meal, 1997

❏ ❏   USA Wa9701 **#1 Bambi (Bambi) -** W Moveable Legs.
                    $2.00-3.00
❏ ❏   USA Wa9702 **#2 Simba (The Lion King) -** W Moveable Legs. $2.00-3.00
❏ ❏   USA Wa9703 **#3 Elliott (Pete's Dragon) -** Green Dragon W Purple Wings. $2.00-3.00
❏ ❏   USA Wa9704 **#4 Dodger (Oliver & Company) -** White Dog W Brn Face/Red Ribbon on Neck. $2.00-3.00
❏ ❏   USA Wa9705 **#5 Princess Aurora (Sleeping Beauty) -** Blue Dress W Gold Locket/Moveable Arms. $2.00-3.00
❏ ❏   USA Wa9706 **#6 Woody (Toy Story) -** Brn Hat/Boots W Gold Shirt/Blue Pants/Moveable Arms/2p. $2.00-3.00
❏ ❏   USA Wa9707 **#7 Donald Duck (The Three Caballeros) -** Brn Mexican Hat W Blanket/Moveable Arms. $2.00-3.00
❏ ❏   USA Wa9708 **#8 Tigger (The Many Adventures of Winnie the Pooh) -** Org Tiger W White Face/Moveable Legs.
                    $2.00-3.00

❏ ❏   Jpn Wa9709 **1998 Calendar.** $5.00-8.00

Comments: Distribution: New Zealand -1997. Seven of the 16 Videos (#2, 3, 4, 7 from Series I, 1996 and #3, 6, 7 from Series II) were combined in several world markets. The USA released two series of eight videos each. Around the world, different numbers of video boxes containing figurines were released in late 1996, 1997, and 1998. Japan released Disney Calendar, Aladdin, Sleeping Beauty, Tigger, and Woody in November, 1997. National Distribution: USA - May 16-June 12, 1997.

USA Wa9701-04

USA Wa9705-08

X-Ray Riders Happy Meal, 1997

❏ ❏   Jpn Xr9701 **Hamburglar on Train -** Clear Vehicle.
                    $4.00-5.00
❏ ❏   Jpn Xr9702 **Birdie on Airplane -** Clear Vehicle.
                    $4.00-5.00
❏ ❏   Jpn Xr9703 **Ronald on Car -** Clear Vehicle. $4.00-5.00
❏ ❏   Jpn Xr9704 **Grimace on Fire Engine -** Clear Vehicle.
                    $4.00-5.00

Comments: National Distribution: Japan - June, 1997; USA - 1999 during clean-up weeks in some markets.

Jpn Xr9701-04

1998

- **14th National Owner/Operator Convention**

- **"Did somebody say McDonald's?"** (Repeated)

- **"Where the World's Best Come Together"** (Olympics 1998)

- **"Made for You"**

- **"Made for You...At the Speed of McDonald's"**

- **"Iam Hungry"** character introduced (computer animated)

Action Man/My Little Pony Happy Meal, 1998

▨ ❏   Uk Ac9801 **GI Man Crawling -** wearing green.
                    $3.00-4.00

Uk Ac9801

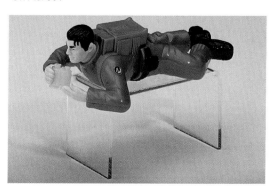

Uk Ac9802 **GI Man Rappelling Down a Rope**.
$3.00-4.00

Uk Ac9803 **GI Man in Astronaut Suit** - gray base.
$3.00-4.00

Uk Ac9804 **GI Man on Black Transfer Line**. $3.00-4.00

Uk My9801 **Little Pony** - pink on blue stand. $4.00-5.00

Uk My9802 **Little Pony** - blue on pink mirror stand.
$4.00-5.00

Uk My9803 **Little Pony** - pink with blue mane. $4.00-5.00

Uk My9804 **Little Pony** - white with lavender mane.
$4.00-5.00

Comments: Distribution: UK - 1998.

Uk Ac9801-04. Insert Card.

Uk My9801-04. Insert Card.

Animal Kingdom Happy Meal, 1998

USA An9801 **#1 Triceratops** - Grey W 2 White Horns Figurine.
$2.00-3.00

USA An9802 **#2 Toucan** - Yel/Red/Grn Beak Bird Figurine.
$2.00-3.00

USA An9803 **#3 Gorilla & Baby** - Blk Gorilla Holding Baby on Brn Pole Figurine $2.00-3.00

USA An9804 **#4 Elephant** - Grey Elephant W Tusks Figurine.
$2.00-3.00

USA An9805 **#5 Ring Tail Lemur** - Brn W Wht Head & Blk Striped Tail Figurine. $2.00-3.00

USA An9806 **#6 Dragon** - Purp Dragon W Lg Purple/Blk Wings.
$2.00-3.00

USA An9807 **#7 Iguanodon** - Turq Iguana Figurine.
$2.00-3.00

USA An9808 **#8 Zebra** - Blk/Wht Striped Figurine.
$2.00-3.00

USA An9809 **#9 Lion** - Yellowish/Brn Lion Figurine.
$2.00-3.00

USA An9810 **#10 Cheetah** - Lt Yellow/Blk Cheetah Figurine.
$2.00-3.00

USA An9811 **#11 Crocodile** - Greenish Crocodile W Open Mouth.
$2.00-3.00

USA An9812 **#12 Rhino** - Grey Rhino W 2 Horns.
$2.00-3.00

USA An9813 **#13 Tortoise - Wal-Mart Special** - Greenish Turtle.
$4.00-5.00

Comments: National Distribution: USA - April 24 - May 21, 1998.

USA An9801-04

USA An9805-08

USA An9809-12

Animal Kingdom Disney Happy Meal, 1998

❏ ❏   Zea Ki9801 **Donald With Ship** - W Discovery River Back-
drop. $3.00-4.00
❏ ❏   Zea Ki9802 **Goofy With Blue Dinosaur** - W Dinoland USA
Backdrop. $3.00-4.00
❏ ❏   Zea Ki9803 **Mickey With Green Jeep** - W Kilimanjaro Sa-
fari Backdrop. $3.00-4.00
❏ ❏   Zea Ki9804 **Minnie With Gorilla** - W Gorilla Falls Explora-
tion Trails Backdrop. $3.00-4.00

Comments: Distribution: New Zealand - June 27, 1998.

Zea Ki9801-04

Animal Kingdom toy available only at Walt
Disney World, Orlando, Florida, USA.

Zea Ki9801-04

USA An9826

Japanese McDonald's
advertisement.

1998

Barbie/Hot Wheels IX Happy Meal, 1998

Barbie Dolls:
- ❏ ❏ USA Ba9801 **Teen Skipper -** Wearing Green Overalls On Pink Heart Shaped Base. $2.00-4.00
- ❏ ❏ USA Ba9802 **Nineties Barbie -** Wearing Denim Jeans and Jacket W Red Shirt On Red Round Base. $2.00-4.00
- ❏ ❏ USA Ba9803 **Eating Fun Kelly -** Baby in Blue/White Highchair. $2.00-4.00
- ❏ ❏ USA Ba9804 **Bead Blast Christie -** Pink/White Striped Dress W Beads In Hair. $2.00-4.00

Hot Wheels:
- ❏ ❏ Hw9805 **Ronald NASCAR #94 -** Red Car W Ronald's Face On Hood. $2.00-4.00
- ❏ ❏ Hw9806 **Mac Tonight Car #94 -** Blue W Mac Tonight On Hood. $2.00-4.00
- ❏ ❏ Hw9807 **Hot Wheels NASCAR #44 -** Blue Car W Hot Wheels On Hood. $2.00-4.00
- ❏ ❏ Hw9808 **50th Anniversary Car #94 -** Silver W 50th Anniversary On Hood. $2.00-4.00

Comments: National Distribution: USA - August 14-September 10, 1998.

Ger Ba9801-04. Insert Card.

USA Ba9801-04

USA Hw9805-08

Beauty & the Beast Enchanted Christmas Happy Meal, 1998

- ❏ ❏ Uk Be9801 **Belle Beauty -** In Red/White Dress. $5.00-6.00
- ❏ ❏ Uk Be9802 **Forte -** Yellow. $4.00-6.00
- ❏ ❏ Uk Be9803 **Cogsworth -** Clock. $4.00-6.00
- ❏ ❏ Uk Be9804 **Beast -** W Purple Cape. $4.00-6.00

Comments: Distribution: UK - November/December, 1998.

Uk Be9801, 04

Uk Be9801-04

Beauty & the Beast Happy Meal, 1998

	❏	Bra Be9801 **Mrs. Potts** - Yellow.	$5.00-7.00
	❏	Bra Be9802 **Belle Beauty** - Red.	$5.00-7.00
	❏	Bra Be9803 **Beast** - Purple.	$5.00-7.00
	❏	Bra Be9804 **Forte** - Gold.	$5.00-7.00

Comments: Distribution: Brazil - 1998.

Bra Be9801-04

Jpn Be9801-06

Bug's Life Happy Meal, 1998

❏	❏	USA Bu9801 **#1 Dim** - Blue Beetle Bug W Blue Shell.	$3.00-4.00
❏	❏	USA Bu9802 **#2 Rosie** - Purple Six Legged Spider.	$3.00-4.00
❏	❏	USA Bu9803 **#3 Dot** - Purple Bug On Blue Mushroom.	$3.00-4.00
❏	❏	USA Bu9804 **#4 Flik** - Purple Bug W Green Backpack On Green Leaf.	$3.00-4.00
❏	❏	USA Bu9805 **#5 Francis** - Grey, Black & Red Beetle Standing Up.	$3.00-4.00
❏	❏	USA Bu9806 **#6 Heimlich** - Green Snail.	$3.00-4.00
❏	❏	USA Bu9807 **#7 Hopper** - Tan Grasshopper.	$3.00-4.00
❏	❏	USA Bu9808 **#8 Atta** - Purple Fly On Green Leaf.	$3.00-4.00

Comments: National Distribution: USA - November 20-December 17, 1998; France - 1999 (all eight toys).

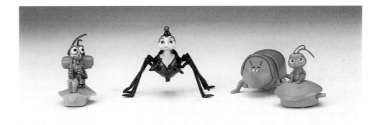

USA Bu9801-04

Beauty & the Beast Enchanted Christmas Happy Meal, 1998

❏	❏	Jpn Be9801 **Disney Video Calendar**.	$4.00-6.00
❏	❏	Jpn Be9802 **Forte**.	$4.00-6.00
❏	❏	Jpn Be9803 **Cogsworth**.	$4.00-6.00
❏	❏	Jpn Be9804 **Angelique with Christmas Tree**.	$4.00-6.00
❏	❏	Jpn Be9805 **Mrs. Potts**.	$4.00-6.00
❏	❏	Jpn Be9806 **Belle Beauty & the Beast**.	$5.00-6.00

Comments: Distribution: Japan - December 1998.

USA Bu9805-08

1998

Bug's Life Insert Card.

Fra Bu9826. French Bug's Life Happy Meal Display.

Characters McDonaldland Brazil Happy Meal, 1998

☐ ☐   Bra Ch9801 **Ronald** - Red Walker/2p. $10.00-15.00
☐ ☐   Bra Ch9802 **Grimace** - Purple Walker/2p. $10.00-15.00
☐ ☐   Bra Ch9803 **Birdie** - Yellow Walker/2p. $10.00-15.00
☐ ☐   Bra Ch9804 **Hamburglar** - Orange Walker/2p.
                                                    $10.00-15.00

Comments: Distribution: Brazil - 1998. Characters resemble the older design style characterization.

USA Bu9826

Bra Ch9801-04

***Bu9826

Characters McDonaldland Far East Happy Meal, 1998

❏ ❏   Sin Ch9801 **Ronald** - Windup. $5.00-7.00
❏ ❏   Sin Ch9802 **Grimace -** Windup. $5.00-7.00
❏ ❏   Sin Ch9803 **Birdie** - Windup. $5.00-7.00
❏ ❏   Sin Ch9804 **Hamburglar** - Windup. $5.00-7.00

Comments: Distribution: Far East, Singapore - 1998. Characters resemble the newer '90s design style characterization.

Sin Ch9802-04

Characters McDonaldland Vehicles Brazil Happy Meal, 1998

❏ ❏   Bra Fl9801 **Ronald in Flying Shoe** - Red W Green Stand. $5.00-7.00
❏ ❏   Bra Fl9802 **Grimace in Airplane** - Purple W Dark Green Stand. $5.00-7.00
❏ ❏   Bra Fl9803 **Birdie in Flying Bird** - Pink W Blue Stand. $5.00-7.00
❏ ❏   Bra Fl9804 **Hamburglar in Airplane -** Orange W Yellow Stand. $5.00-7.00

Comments: Distribution: Brazil - 1998. Characters resemble the newer '90s design style characterization.

Bra Fl9801-04

Disney Video Favorites Happy Meal, 1998

❏ ❏   USA Di9801 **The Spirit of Mickey on Video Box.** $3.00-5.00
❏ ❏   USA Di9802 **Lady and the Tramp on Video Box.** $3.00-5.00
❏ ❏   USA Di9803 **Pocahontas: Journey to a New World on Video Box.** $3.00-5.00
❏ ❏   USA Di9804 **Mary Poppins on Video Box -** Penguin on Box. $3.00-5.00
❏ ❏   USA Di9805 **The Black Cauldron on Video Box -** Animal Figure on Box. $3.00-5.00
❏ ❏   USA Di9806 **Flubber on Video Box.** $3.00-5.00

Comments: National Distribution: USA - September 11-October 1, 1998.

USA Di9801-03

USA Di9804-06

Disneyland Paris Costume Happy Meal, 1998

❏ ❏   Fra Pa9801 **Mickey's Ears -** Black Cap Sectional Piece. $10.00-12.00
❏ ❏   Fra Pa9802 **Mickey's Left Hand -** Glove. $6.00-8.00
❏ ❏   Fra Pa9803 **Mickey's Right Hand -** Glove. $6.00-8.00
❏ ❏   Fra Pa9804 **Mickey's Face Mask -** Face Sectional Piece. $6.00-8.00

Comments: Distribution: France - 1998. All four pieces formed Mickey's costume, to be worn by children. The gloves and black cap with Mickey's ears were made of cloth, the mask was made of plastic.

Fra Pa9801-04

Fra Pa9801-04

Disneyland Paris Insert Card.

Flubber Happy Meal, 1998

☐ ☐ Zea FI9801 **Flubber Tank -** Water Tank Style W Lump Of Green Flubber On Inside. $3.00-4.00

☐ ☐ Zea FI9802 **Weber -** Gray Robot W Arms & Base, Windup. $3.00-4.00

☐ ☐ Zea FI9803 **Weebo/Moose -** Yellow Robot W Arm. $3.00-4.00

☐ ☐ Zea FI9804 **T-Bird/Action/ Thunderbird Car -** Red With **Robin Williams Driving.** $5.00-8.00

☐ ☐ Zea FI9805 **Chemistry Set.** $4.00-5.00

Comments: Distribution: Japan - 1998; Parts of Europe distributed FI9805; France - 1998 (#1-4); UK - February, 1998; Australia - January, 1998.

Flubber Set with four toys.

German Flubber Set with five toys.

Ger FI9726. German Flubber Poster.

Flubber Happy Meal Box.

Fra FI9826

French Flubber Happy Meal Box.

French Counter Card.

Fra Fl9864. French Translite.

Zea Fl9801-05

France '98 World Cup Soccer Happy Meal, 1998

☐ ☐ Aus Fr9801 **Soccer Ball -** World Cup Logo. $3.00-4.00
☐ ☐ Aus Fr9802 **Sipper Cup -** World Cup Logo. $3.00-4.00
☐ ☐ Aus Fr9803 **Baseball/Handball -** World Cup Logo.
$3.00-4.00

Comments: Distribution: Australia - July 1998.

Aus Fr9801-03

USA Ha9801-03

USA Ha9804-06

Halloween '98 Ronald and Pals Happy Meal, 1998

☐ ☐ USA Ha9801 **#1 Iam Hungry -** Green Character W Purple
Witch's Face/2p. $3.00-4.00
☐ ☐ USA Ha9802 **#2 Birdie -** Pink Character W Blk Cat's Face/
Blue Bow/2p. $3.00-4.00
☐ ☐ USA Ha9803 **#3 Grimace -** Purple Character W Orange
Pumpkin Face/2p. $3.00-4.00

☐ ☐ USA Ha9804 **#4 McNugget Buddy -** Brown Chicken Nug-
get W White Ghost Face.2p. $3.00-4.00
☐ ☐ USA Ha9805 **#5 Ronald -** Red Ronald W Blue Spooky Face/
Purple Hat/2p. $3.00-4.00
☐ ☐ USA Ha9806 **#6 Hamburglar -** Black Character W Green
Ghost Face W Purple Spider On Tongue. $3.00-4.00

Comments: National Distribution: USA - October 9-October 29,
1998.

Hercules II Happy Meal, 1998

☐ ☐ USA He9801 **#1 Zeus Football -** Orange Mini Football W
Zeus' Face In White. $2.00-4.00
☐ ☐ USA He9802 **#2 Hades Stopwatch -** Blue/Grey/Yellow Plas-
tic Clock Shaped Stopwatch. $2.00-4.00
☐ ☐ USA He9803 **#3 Hercules Sports Bottle -** Blue Column
Shaped Bottle W Hercules Picture. $2.00-4.00

USA He9801-03

☐ ☐ USA He9804 **#4 Eyes of Fates Foot Bag -** White Bean Bag
W Eye Imprint. $2.00-4.00

☐ ☐ USA He9805 **#5 Pegasus Whistling Discus -** Silver Disc.
$2.00-4.00

☐ ☐ USA He9806 **#6 Pain and Panic Sound Baton -** Purple
Column Shaped Sound Stick. $2.00-4.00

☐ ☐ USA He9807 **#7 Hercules Medal -** Silver/Black Locket
Hercules. $2.00-4.00

☐ ☐ USA He9808 **#8 Phil Megaphone -** Purple/Blue/Beige
Megaphone. $2.00-4.00

Comments: National Distribution: USA - January 30-February 26,
1998.

USA He9804-08

Le Crazy Happy Meal, 1998

☐ ☐ Fra Lc9801 **Merlin -** Disney Box W Classic Photos.
$3.00-4.00

☐ ☐ Fra Lc9802 **Robin Hood -** Disney Box W Classic Photos.
$3.00-4.00

☐ ☐ Fra Lc9803 **Jungle Book -** Disney Box W Classic Photos.
$3.00-4.00

☐ ☐ Fra Lc9804 **The Aristocats -** Disney Box W Classic Photos.
$3.00-4.00

☐ ☐ Fra Lc9805 **Bernand & Bianca -** Disney Box W Classic Photos. $3.00-4.00

Comments: Distribution: France - 1998. First advertised Happy
Meal FOR THE FAMILY.

Fra Lc9801-05

Fra Lc9801

Fra Lc9802

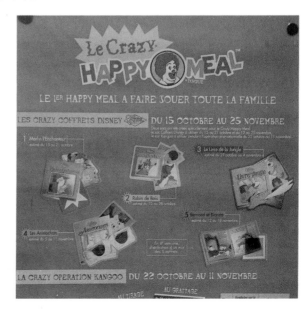

Legend of Mulan Happy Meal, 1998

- USA Mu9801 **#1 Mulan -** Girl With Purple Apron Sporting Black And Green Armour/2p. $2.00-3.00
- USA Mu9802 **#2 Cri-Kee -** Purple Cricket. $2.00-3.00
- USA Mu9803 **#3 Khan -** Black Horse W Yellow Saddle. $2.00-3.00
- USA Mu9804 **#4 Shan-Yu -** Samuri Warrior W Sword. $2.00-3.00
- USA Mu9805 **#5 Little Brother -** White Dog. $2.00-3.00
- USA Mu9806 **#6 Shang-Li -** Male Samurai Warrior W Green Jacket. $2.00-3.00
- USA Mu9807 **#7 Mushu -** Dragon W Gold Gong. $2.00-3.00
- USA Mu9808 **#8 Chien-Po & Ling, Yao -** Summo Wrestler W Figure On Rope/2p. $2.00-3.00

Comments: Distribution: USA - June 19-July 16, 1998; New Zealand - 1998.

USA Mu9801-05

USA Mu9806-08

Little Mermaid III Happy Meal, 1998

- Bel Li9801 **Ariel on Log -** Windup W Disappearing Fork. $3.00-5.00
- Bel Li9802 **Glut the Shark -** In Oceanlike Box, Windup. $3.00-5.00
- Bel Li9803 **Scuttle Eureka the Seagull -** Floating On Back, Windup. $3.00-5.00
- Bel Li9804 **Sebastian -** Red Crab, Windup. $3.00-5.00

Comments: Distribution: Belgium - July, 1998. Noteworthy is the production location of toys distributed: Ariel, Glut, and Scuttle made in China; Sebastian made in Vietnam; paper insert printed in Hong Kong. This Happy Meal illustrates the global nature of McDonald's toys.

Bel Li9801-04. Little Mermaid Insert Card.

Marvel Superheroes Happy Meal, 1998

- Aus Ma9801 **Wolverine.** $4.00-5.00
- Aus Ma9802 **Spiderman.** $4.00-5.00
- Aus Ma9803 **The Thing.** $4.00-5.00
- Aus Ma9804 **Silver Surfer.** $4.00-5.00

Comments: Distribution: Australia - September, 1998.

Aus Ma9801-04

McDonaldland Soccer TEAM 3000 Happy Meal, 1998

- ❏ ❏ Jpn So9801 **Ronald With Pink Soccer Ball.** $4.00-5.00
- ❏ ❏ Jpn So9802 **Hamburglar With Blue Soccer Ball.**
  $4.00-5.00
- ❏ ❏ Jpn So9803 **Grimace With Orange Soccer Ball.**
  $4.00-5.00
- ❏ ❏ Jpn So9804 **Birdie With Green Soccer Ball.** $4.00-5.00

Comments: Distribution: Japan - July, 1998

McKit Happy Meal, 1998

- ❏ ❏ Jpn Ki9801 **Ronald With Paper, Notebox.** $5.00-7.00
- ❏ ❏ Jpn Ki9802 **Grimace With Ruler.** $5.00-7.00
- ❏ ❏ Jpn Ki9803 **Birdie With Crayons, Sharpener.**
  $5.00-7.00
- ❏ ❏ Jpn Ki9804 **Hamburglar With Scissors.** $5.00-7.00

Comments: Distribution: Japan - 1998. All four toys snap together to form one large house.

Jpn Ki9801-04

Japanese advertising for McKit Happy Meal.

McNugget Buddies Brazil Happy Meal, 1998

- ❏ ❏ Bra Nu9801 **McNugget - Red Frankenstein.** $5.00-7.00
- ❏ ❏ Bra Nu9802 **McNugget - White Ghost.** $5.00-7.00
- ❏ ❏ Bra Nu9803 **McNugget - Black Dracula.** $5.00-7.00
- ❏ ❏ Bra Nu9804 **McNugget - Red/Green Werewolf.**
  $5.00-7.00

Comments: Distribution: Brazil - 1998.

Bra Nu9801-04

Mickey's Adventure Happy Meal, 1998

- ❏ ❏ Jpn Ad9801 **Donald's Boat House -** In Blue, White & Brown.
  $5.00-6.00
- ❏ ❏ Jpn Ad9802 **Minnie's Cottage -** In Blue, White & Green.
  $5.00-6.00
- ❏ ❏ Jpn Ad9803 **Goofy's Bounce House -** In Red, Green & White.
  $5.00-6.00
- ❏ ❏ Jpn Ad9804 **Mickey's Adventure House -** In Red, Tan & Gray.
  $5.00-6.00

Comments: Distribution: Japan - May, 1998.

Jpn Ad9801-04

Mighty McDinos Happy Meal, 1998

❑ ❑ Zea Mi9801 **Ankylosaurus Inside Shell** - Soft Stuffed Fig In
2p Shell. $2.00-3.00
❑ ❑ Zea Mi9802 **Pterodactyl Inside Shell** - Soft Stuffed Fig In
2p Shell. $2.00-3.00
❑ ❑ Zea Mi9803 **Triceratops Inside Shell** - Soft Stuffed Fig In
2p Shell. $2.00-3.00
❑ ❑ Zea Mi9804 **Stegosaurus Inside Shell** - Soft Stuffed Fig In
2p Shell. $2.00-3.00

Comments: Distribution: New Zealand - 1998; Brazil - 1998. These
toys are plush toys that have a hard, plastic, removable shell.

My Little Pony/Transformers Beast Wars II Happy Meal, 1998

My Little Pony:
❑ ❑ USA My9801 **#1 Ivy -** Turquoise Pony W Pink Tail W Purp
Mane. $4.00-5.00
❑ ❑ USA My9802 **#2 Sundance -** Pink Pony W Pink Tail W Yel-
low Sunburst On Rump, Blue/Pink Mane. $4.00-5.00
❑ ❑ USA My9803 **#3 Light Heart -** White Pony W Pink Heart
On Rump, Purple/Beige Tail. $4.00-5.00

USA My9826

USA My9801-03

USA My9804-06

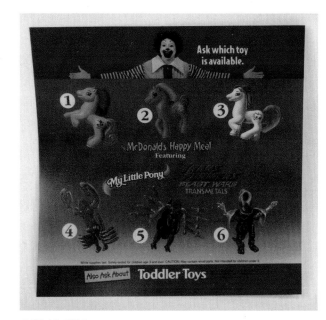

USA My9826

Transformers:
❑ ❑ USA My9804 **#4 Scorponok -** Purp/Blue/Grey Crablike Fig-
ure W Crab Claws. $4.00-5.00
❑ ❑ USA My9805 **#5 Blackarachnia -** Purple Body/Blue Mask/
Org Claws Forming Beetlelike Figure. $4.00-5.00
❑ ❑ USA My9806 **#6 Dinobot -** Blue/Silver/Green Serpentlike
Figure W Green Head Mask. $4.00-5.00

Comments: National Distribution: USA - March 27-April 17, 1998.

Nagano/Speed Sliders Happy Meal, 1998

❏	❏	Jpn Na9801 **Ronald in Bobsled**.	$4.00-5.00
❏	❏	Jpn Na9802 **Birdie as Luger**.	$4.00-5.00
❏	❏	Jpn Na9803 **Grimace as Ski Jumper**.	$4.00-5.00
❏	❏	Jpn Na9804 **Hamburglar as Hockey Player**.	$4.00-5.00

Comments: Distribution: Nagano, Japan for the Olympic Games.

Jpn Na9801-04

Nagano Japan Finish Line Happy Meal, 1998

❏ ❏ Jpn Na9805 **Ronald in Bobsled - W Finish Line Sign**.
$5.00-6.00

❏ ❏ Jpn Na9806 **Birdie as Luger - W Finish Line Sign**.
$5.00-6.00

❏ ❏ Jpn Na9807 **Grimace as Ski Jumper - W Finish Line Sign**.
$5.00-6.00

❏ ❏ Jpn Na9808 **Hamburglar as Hockey Player - W Finish Line Sign**.
$5.00-6.00

Comments: Distribution: Nagano, Japan for the Olympic Games. Finish line reads, "Nagano, Japan." A second set was distributed without the Nagano, Japan sign on finish line.

Jpn Na9805-08. Display.

Nagano Japan Happy Meal, 1998

❏	❏	Jpn Na9809 **Bobsleders**.	$5.00-6.00
❏	❏	Jpn Na9810 **Skiers**.	$5.00-6.00
❏	❏	Jpn Na9811 **Luge - Tape Dispenser**.	$5.00-6.00
❏	❏	Jpn Na9812 **Downhill - Stapler**.	$5.00-6.00

Comments: Distribution: Nagano, Japan for the Olympic Games.

Jpn Na9809-12

Peter Pan Build a Scene Happy Meal, 1998.

❏ ❏ Aus Pe9801 **Scene: Big Ben & London Bridge**.
$3.00-5.00

❏	❏	Aus Pe9802 **Scene: Pirate Ship**.	$3.00-5.00
❏	❏	Aus Pe9803 **Scene: Island**.	$3.00-5.00
❏	❏	Aus Pe9804 **Scene: Indian Village**.	$3.00-5.00

Comments: Distribution: Australia - June 1998.

Aus Pe9801-04

1998

Peter Pan Happy Meal, 1998

- ❏ ❏ USA Pe9801 **#1 Peter Pan Glider** - Peter Pan Fig W Green Clothes Attached To White Wings. $3.00-5.00
- ❏ ❏ USA Pe9802 **#2 Tick Tock Crocodile Compass** - Crocodile/Green W Compass In Fliptop Mouth. $3.00-5.00
- ❏ ❏ USA Pe9803 **#3 Captain Hook Spyglass** - Red Spyglass W Capt. Hook's Head/Arm On Handle. $3.00-5.00
- ❏ ❏ USA Pe9804 **#4 Wendy & Michael Magnifier** - Wendy & Michael Fig Imprinted On Green Magnifier. $3.00-5.00
- ❏ ❏ USA Pe9805 **#5 Peter Pan Activity Tool** - Brown Knifelike Tool W Green Crocodile Tool. $3.00-5.00
- ❏ ❏ USA Pe9806 **#6 Tinker Bell Lantern** - Tinker Bell Inside Red/Yellow/Blue Sm Lantern. $3.00-5.00
- ❏ ❏ USA Pe9807 **#7 Smee Light** - Man In Brown Barrel W Red Flashlight Effect. $3.00-5.00

Comments: National Distribution: USA - February 27-March 26, 1998.

Pumucki Happy Meal, 1998

Toys: Series I

- ❏ ❏ Ger Pu9801 **Pumucki with 10th Anniversary Sign.** $5.00-7.00
- ❏ ❏ Ger Pu9802 **Pumucki on Bed.** $5.00-7.00
- ❏ ❏ Ger Pu9803 **Pumucki with Trumpet.** $5.00-7.00
- ❏ ❏ Ger Pu9804 **Pumucki Laying Down Painting, Creating Sign.** $5.00-7.00
- ❏ ❏ Ger Pu9805 **Pumucki Wrapping Present.** $5.00-7.00

Toys: Series II

- ❏ ❏ Ger Pu9806 **Pumucki with 10th Anniversary Cake.** $5.00-7.00
- ❏ ❏ Ger Pu9807 **Pumucki with Soda/Drink.** $5.00-7.00
- ❏ ❏ Ger Pu9808 **Pumucki with Fries.** $5.00-7.00
- ❏ ❏ Ger Pu9809 **Pumucki with Hamburger.** $5.00-7.00
- ❏ ❏ Ger Pu9810 **Pumucki with 10th Anniversary Junior Tute Happy Meal Bag Holding Food Items.** $5.00-7.00

USA Pe9801-03

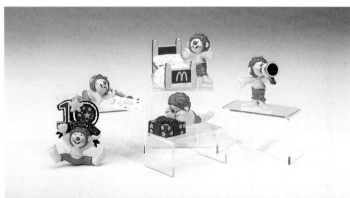

Ger Pu9801-05

USA Pe9804-07

USA Pe9826

Ger Pu9806-10

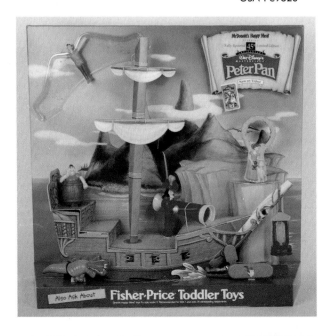

Toy: Club Member Figurines
- ❏ ❏ Ger Pu9811 **Pumucki with Jr. Club sign**. $10.00-15.00

Comments: Distribution: Germany - June, 1998. Two series highlighted this promotion followed by a special figurine given out at birthday parties to the "Birthday Child" and/or through special promotional offers.

Ger Pu9811

104

German Pumucki Insert Card.

Ger Pu9826

Recess Happy Meal, 1998

- ❏ ❏ USA Re9801 **#1 TJ -** Boy W Yellow Hockey Stick/Red Ball/2p. $2.00-3.00
- ❏ ❏ USA Re9802 **#2 Spinelli -** Girl W Green Bat/White Ball/2p. $2.00-3.00
- ❏ ❏ USA Re9803 **#3 Vince - African American** Boy W #1 On Shirt/Orange Basketball/2p. $2.00-3.00
- ❏ ❏ USA Re9804 **#4 Gretchen -** Girl W Glasses and Tennis Racket/Green Ball/2p. $2.00-3.00

- ❏ ❏ USA Re9805 **#5 School Teacher Miss Finster -** Teacher In Yellow Dress/Purple Ball/2p. $2.00-3.00
- ❏ ❏ USA Re9806 **#6 Gus -** Boy W Glasses and Gold Club/Yellow Ball/2p. $2.00-3.00
- ❏ ❏ USA Re9807 **#7 Mikey -** Boy W White Shirt/Green Ball/2p. $2.00-3.00

Comments: National Distribution: USA - December 25, 1998-January 21, 1999.

USA Re9801-04

USA Re9805-07

German Pumucki Junior Tute Flags.

1998

Sesame Street Live Placemats Happy Meal, 1998

❏ ❏ Aus Se9801 **Placemat: Big Bird**. $3.00-4.00
❏ ❏ Aus Se9802 **Placemat: Elmo**. $3.00-4.00
❏ ❏ Aus Se9803 **Placemat: Oscar the Grouch**. $3.00-4.00
❏ ❏ Aus Se9804 **Placemat: Cookie Monster**. $3.00-4.00

Comments: Distribution: Australia - January 1998.

Aus Se9801-04

Simba's Pride Lion King II Happy Meal, 1998

❏ ❏ USA Si9801 **#1 Kovu** - Stuffed Animal Brn/Blk W Yel/Org Eyes. $2.00-3.00
❏ ❏ USA Si9802 **#2 Zazu** - Stuffed Birdlike Animal Blue/Wht/ Org. $2.00-3.00
❏ ❏ USA Si9803 **#3 Timon** - Stuffed Animal Tan/Brn. $2.00-3.00
❏ ❏ USA Si9804 **#4 Kiara** - Stuffed Animal Tan W Purple Nose. $2.00-3.00
❏ ❏ USA Si9805 **#5 Pumbaa** - stuffed Animal Tan/Pnk Nose. $2.00-3.00
❏ ❏ USA Si9806 **#6 Ziro** - Stuffed Animal Brn/Blk/Tan W Long Body/Nose. $2.00-3.00
❏ ❏ USA Si9807 **#7 Rafiki** - Stuffed Animal Purple/Whit W Wht Beard. $2.00-3.00
❏ ❏ USA Si9808 **#8 Simba** - Stuffed Lion Animal Tan/Brn Fringe Hair/Beard. $2.00-3.00

Comments: National Distribution: USA - 1998; France - 1999; all eight toys.

USA Si9801-04

USA Si9805-08

Smurfs Happy Meal, 1998

❏ ❏ Uk Sm9811 **Smurf with Happy Meal Box**. $7.00-10.00

Comments: Distribution: UK - April, 1998. Set of ten included nine Smurfs identical to previous set from Germany (Ger Sm9601-10), except that Smurf carrying Happy Meal replaced Smurf with Numeral 25th Anniversary Celebration sign (Ger Sm9606).

Uk Sm9811

Tamagotchi Happy Meal, 1998

❏ ❏ USA To9801 **#1 Yellow Egg Shaped**. $2.00-3.00
❏ ❏ USA To9802 **#2 Purple Egg Shaped W Green Dog Figure**/2p. $2.00-3.00
❏ ❏ USA To9803 **#3 Green Egg Shaped**. $2.00-3.00
❏ ❏ USA To9804 **#4 Red Egg Shaped**. $2.00-3.00

USA To9801-04

❏ ❏   USA To9805 **#5 Blue Egg Shaped W Yellow Figure**/2p.
$2.00-3.00
❏ ❏   USA To9806 **#6 White Egg Shaped W Black And Red Figure**/2p.
$2.00-3.00
❏ ❏   USA To9807 **#7 Orange Egg Shaped.** $2.00-3.00
❏ ❏   USA To9808 **#8 Blue Egg Shaped.** $2.00-3.00

❏ ❏   USA To9809 **#9 Dark Blue/Lime Green Glowing Tamagotchi.** $5.00-8.00

Comments: National Distribution: USA - July 17-August 13, 1998.

USA Ty9805-08

USA To9805-08

Teenie Beanie Babies II Happy Meal, 1998

❏ ❏   USA Ty9801 **#1 Doby the Doberman -** brown and black.
$5.00-8.00
❏ ❏   USA Ty9802 **#2 Bongo the Monkey -** brown.$4.00-5.00
❏ ❏   USA Ty9803 **#3 Twigs the Girafe -** orange and yellow.
$4.00-5.00
❏ ❏   USA Ty9804 **#4 Inch the Worm -** yellow, orange, green, black and purple. $4.00-5.00

❏ ❏   USA Ty9805 **#5 Pincher the Lobster -** red. $4.00-5.00
❏ ❏   USA Ty9806 **#6 Happy the Hippo -** purple. $4.00-5.00
❏ ❏   USA Ty9807 **#7 Mel the Koala -** grey and white.
$4.00-5.00
❏ ❏   USA Ty9808 **#8 Scoop the Pelican -** large orange beak.
$4.00-5.00

❏ ❏   USA Ty9809 **#9 Bones the Dog -** tan with brown ears.
$4.00-5.00
❏ ❏   USA Ty9810 **#10 Zip the Cat -** black with white paws.
$4.00-5.00
❏ ❏   USA Ty9811 **#11 Waddle the Penguin -** black and white with orange feet. $4.00-5.00
❏ ❏   USA Ty9812 **#12 Peanut the Elephant -** blue.
$4.00-5.00

Comments: National Distribution: USA - May 22-June 18, 1998.

USA Ty9809-12

Underwater Adventure Happy Meal, 1998

❏ ❏   Zea Un9801 **Alvin Explorer -** green. $3.00-4.00
❏ ❏   Zea Un9802 **Nautile Sea Robot -** orange. $3.00-4.00
❏ ❏   Zea Un9803 **Nautilus Submarine -** blue. $3.00-4.00
❏ ❏   Zea Un9804 **Ronald Turtle -** red. $3.00-4.00

Comments: Distribution: New Zealand - 1998.

Village/McDonaldland Village Happy Meal 1994/1993, 1998

❏ ❏   *** Vi9801 **Village House/Blue Roof** - Birdie in Blue Jeep W Sticker Sheet. $1.00-2.00
❏ ❏   *** Vi9802 **Village House/Green Roof** - Fry Kids in Red Fire Truck W Sticker Sheet. $1.00-2.00
❏ ❏   *** Vi9803 **Village House/Red Roof** - Ronald in Red Shoe Car W Sticker Sheet. $1.00-2.00
❏ ❏   *** Vi9804 **Village House/Yellow Roof** - Grimace in Green Engine W Sticker Sheet. $1.00-2.00

❏ ❏   *** Vi9805 **Birdie in Blue Jeep** - W Sticker Sheet.
$1.00-2.00
❏ ❏   *** Vi9806 **Fry Kids in Red Fire Truck** - W Sticker Sheet.
$1.00-2.00
❏ ❏   *** Vi9807 **Ronald in Red Shoe Car** - W Sticker Sheet.
$1.00-2.00
❏ ❏   *** Vi9808 **Grimace in Green Engine** - W Sticker Sheet.
$1.00-2.00

USA Ty9801-04

Generic advertising poster.

Comments: Regional Distribution: USA, Canada - 1998 during clean-up periods; Europe/UK -November-December, 1993; Chile - May, 1994; Japan - June, 1994; Quebec, Canada, Venezuela, New Zealand - March, 1994; Argentina - October, 1994. A second set with only the vehicles was distributed in many markets around the world during clean-up periods. Happy Meal Set in Japan is called "Okosama Set." Note that the roof colors vary according to the countries distributed: red, blue, green, yellow, purple, or pink. China's Set has lighter colors on the roof, ie. pink instead of red, purple instead of green. Village Houses were packaged in Germany and China; not specifically distributed in Germany.

Saudia Arabia, advertising poster.

***Vi9801-04

***Vi9805-08

Winnie the Pooh's Treehouse Most Grand Adventure Happy Meal, 1998

❏	❏	Zea Wt9801 **Eeyore**.	$4.00-5.00
❏	❏	Zea Wt9802 **Rabbit With Monster Tree Trunk**.	
			$4.00-5.00
❏	❏	Zea Wt9803 **Tigger On Top Of Tree**.	$4.00-5.00
❏	❏	Zea Wt9804 **Winnie the Pooh**.	$4.00-5.00

Comments: Distribution: New Zealand - 1998. Four toys form a scene around the tree.

Zea Wt9801-04

Winnie the Pooh's Happy Meal, 1998

❏	❏	*** Wt9801 **Eeyore** - In Green Sectional House.	
			$4.00-5.00
❏	❏	*** Wt9802 **Rabbit** - In Purple Sectional House.	
			$4.00-5.00
❏	❏	*** Wt9803 **Tigger** - In Blue Sectional House.	$4.00-5.00
❏	❏	*** Wt9804 **Winnie the Pooh** - In Pink Sectional House.	
			$4.00-5.00

Comments: Distribution: Far East - 1998. Toys stack two by two.

***Wt9801-03

Winnie the Pooh Plush Happy Meal, 1998

❏	❏	Jpn Wi9801 **Winnie** - W Red Shirt, Plush.	$6.00-8.00
❏	❏	Jpn Wi9802 **Tigger** - Plush.	$6.00-8.00
❏	❏	Jpn Wi9803 **Winnie** - W Blue Night Shirt, Plush.	
			$6.00-8.00
❏	❏	Jpn Wi9804 **Winnie** - W faded Red Shirt & Cap, Plush.	
			$6.00-8.00
❏	❏	Jpn Wi9805 **Eeyore** - Plush.	$6.00-8.00

Comments: Distribution: Japan - 1998.

Jpn Wi9801-05

Winnie the Pooh Happy Meal, 1998

▨	❏	Jpn Wp9801 **Winnie the Pooh** - Red Shirt, Plush.	
			$4.00-6.00
❏	❏	Jpn Wp9802 **Eeyore** - Purple, Plush.	$4.00-6.00
❏	❏	Jpn Wp9803 **Tigger** - Orange & Black, Plush.	$4.00-6.00
❏	❏	Jpn Wp9804 **Piglet** - Pink, Plush.	$4.00-6.00

Distriubtion: Japan - 1998.

Jpn Wp9801-04

Jpn Wp9801-04

Winnie the Pooh Scouts Happy Meal, 1998

❑ ❑ *** Ws9801 **Winnie the Pooh -** as Boy Scout Leader.
$7.00-10.00
❑ ❑ *** Ws9802 **Piglet.** $7.00-10.00
❑ ❑ *** Ws9803 **Gopher.** $7.00-10.00
❑ ❑ *** Ws9804 **Owl.** $7.00-10.00
❑ ❑ *** Ws9805 **Tigger.** $7.00-10.00
❑ ❑ *** Ws9806 **Roo.** $7.00-10.00
❑ ❑ *** Ws9807 **Rabbit.** $7.00-10.00
❑ ❑ *** Ws9808 **Eeyore.** $7.00-10.00

Comments: Distribution: Uncertain, may be prototype toys.

***Ws9801-08

World Cup Happy Meal, 1998

❑ ❑ Bel Wo9801 **Speed Ball -** Measures Speed Of Throw.
$5.00-7.00
❑ ❑ Bel Wo9802 **Watch LCD.** $3.00-5.00
❑ ❑ Bel Wo9803 **Frisbee -** Soft, Cloth W World Cup Logo.
$3.00-5.00
❑ ❑ Bel Wo9804 **Windsock -** Mini Kite On String/Pole.
$3.00-5.00
❑ ❑ Bel Wo9805 **Pinball Rectangle Game.** $8.00-10.00

Comments: Distribution: Belgium - 1998, all five premiums; parts of Europe, including Austria - 1998, only toys #1-4; UK - June, 1998, toys #1-4.

Bel Wo9801-05

Fra Wo9801-04

World Wildlife Federation Happy Meal, 1998

❏ ❏   Zea Wi9801 **Kiwi Bird -** Stuffed. $2.00-3.00
❏ ❏   Zea Wi9802 **Black & White Penguin -** Stuffed.
                                   $2.00-3.00
❏ ❏   Zea Wi9803 **White Lamb/Sheep -** Stuffed W Blue Face.
                                   $2.00-3.00
❏ ❏   Zea Wi9804 **Whale -** Stuffed. $2.00-3.00

Comments: Distribution: New Zealand - 1998.

Zea Wi9801-04

1999

• McDonald's opens 25,000th restaurant in Chicago, Illinois

• Snoopy, Furby, Teenie Beanie Babies promotions soar!

• Happy Meal toy sets reach 80 toys per month in some markets.

• Toy sets combine for **COLLECT AND BUILD PROGRAM**

• Toy sets contain a special toy section that forms a different premium when assembled.

Amusement/McAmusement Park Happy Meal, 1999

❏ ❏   Zea Am9901 **Ronald in 360 Degree Loop Ride.**
                                   $4.00-5.00
❏ ❏   Zea Am9902 **Grimace in Windup Pirate Ship.**
                                   $4.00-5.00
❏ ❏   Zea Am9903 **Birdie in Speedboat.** $4.00-5.00
❏ ❏   Zea Am9904 **Hamburglar in Parachute Drop.**
                                   $4.00-5.00

Comments: Distribution: New Zealand - July, 1999.

Zea Am9901-04

Barbie/Hot Wheels Happy Meal, 1999

❏ ❏   Fra Ba9901 **Barbie Princesse Velours -** Red Paper Doll Set.
                                   $5.00-10.00
❏ ❏   Fra Ba9902 **Barbie Barbie Meches Bleues -** Blue Paper
Doll Set. $5.00-10.00
❏ ❏   Fra Ba9903 **Barbie Fantaisie Fruits -** Pink Paper Doll Set.
                                   $5.00-10.00
❏ ❏   Fra Ba9904 **Barbie Decalcomanies Les Papillons -** Pink
Halter Top W Blue Skirt Outfit. $5.00-10.00

❏ ❏   Fra Hw9901 **Truck -** Yellow W Red Cab & HOT WHEELS
Imprinted On Side. $4.00-5.00
❏ ❏   Fra Hw9902 **Car -** Blue W Gold Ramp. $4.00-5.00
❏ ❏   Fra Hw9903 **Car -** Green W Red Ramp. $4.00-5.00
❏ ❏   Fra Hw9904 **Car -** Red W Blue Tunnel Ramp. $4.00-5.00

Comments: Distribution: France - 1999.

Fra Ba9901-04

Fra Ba9901

1999

Fra Ba9901

Fra Ba9901-04. Paper doll punchout booklet.

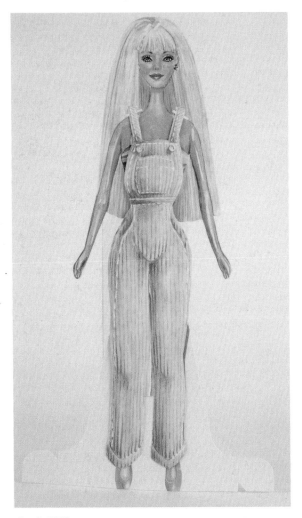

Fra Ba9902

Fra Ba9926. French Barbie Display.

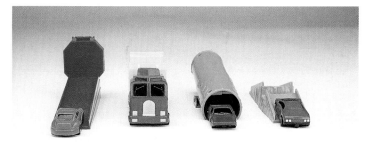

Fra Hw9901-04

Fra Hw9901-04. Insert Card.

Disney Parade Happy Meal, 1999

		Uk Di9901 **Mickey on Stand.**	$3.00-5.00
		Uk Di9902 **Donald on Stand.**	$3.00-5.00
		Uk Di9903 **Goofy on Stand.**	$3.00-5.00
		Uk Di9904 **Chip n Dale on Stand.**	$3.00-5.00
		Uk Di9905 **Minnie on Stand.**	$3.00-5.00
		Uk Di9906 **Daisy on Stand.**	$3.00-5.00
		Uk Di9907 **Dumbo on Stand.**	$3.00-5.00
		Uk Di9908 **Peter Pan on Stand.**	$3.00-5.00
		Uk Di9909 **Pinocchio on Stand.**	$3.00-5.00
		Uk Di9910 **Alice on Stand.**	$3.00-5.00
		Uk Di9911 **Rabbit on Stand.**	$3.00-5.00
		Uk Di9912 **Cheshire Cat on Stand.**	$3.00-5.00
		Uk Di9913 **Sleeping Beauty on Stand.**	$3.00-5.00
		Uk Di9914 **Snow White on Stand.**	$3.00-5.00
		Uk Di9915 **Captain Hook on Stand.**	$3.00-5.00
		Uk Di9916 **Woody on Stand.**	$3.00-5.00
		Uk Di9917 **Buzz Light on Stand.**	$3.00-5.00
		Uk Di9918 **Aladdin on Stand.**	$3.00-5.00
		Uk Di9919 **Pluto on stand.**	$3.00-5.00
		Uk Di9920 **Disney's Enchanted Castle on Stand.**	$5.00-8.00

Comments: Distribution: UK - April 1999.

Bug's Life Puzzle Happy Meal, 1999

		Aus Bu9901 **Puzzle Set -** 2 Picture Puzzles.	$2.00-3.00
		Aus Bu9902 **Puzzle Set -** 2 Picture Puzzles.	$2.00-3.00
		Aus Bu9903 **Puzzle Set -** 2 Picture Puzzles.	$2.00-3.00
		Aus Bu9904 **Puzzle Set -** 2 Picture Puzzles.	$2.00-3.00

Comments: Distribution: Australia - January, 1999.

Uk Di9901-20

Aus Bu9901-04

Uk Di9901-05

Uk Di9906-10

Uk Di9911-15

Uk Di9916-20

Disneyland Paris Build A Mickey Happy Meal, 1999

❏ ❏ Fra Pa9901 **Mickey's Head/Ears - Viewer Sectional Piece.** $8.00-10.00

❏ ❏ Fra Pa9902 **Mickey's Arms - Squirt Gun Sectional Piece.** $6.00-8.00

❏ ❏ Fra Pa9903 **Mickey's Pants - Binoculars Sectional Piece.** $6.00-8.00

❏ ❏ Fra Pa9904 **Mickey's Legs - Stamper Sectional Piece.** $6.00-8.00

Comments: Distribution: France, Australia, Germany - April, 1999, 19" Mickey. Australian Mickey is smaller than the French or German Mickey (14").

Fra Pa9901-04

Fra Pa9901-04, 19" Mickey.

Extreme Sports Happy Meal, 1999

❑	❑	Zea Ex9901 **Hamburglar Motorcycle -** Yel Cyle W Blue Wheels, Red & Black Riding Outfit.	$4.00-5.00
❑	❑	Zea Ex9902 **Birdie Jet Ski -** Pink/Purple Jet Ski W Org/Pink Outfit.	$4.00-5.00
❑	❑	Zea Ex9903 **Grimace Roller Blading with Ramp -** On Green Blades W Blue Ramp/2p.	$4.00-5.00
❑	❑	Zea Ex9904 **Ronald on Skateboard -** Purple/Org Skateboard W Yel/Org Outfit.	$4.00-5.00

Comments: Distribution: New Zealand - 1999.

Zea Ex9901-04

Fra Pa9926. French Display.

Farm Friends Happy Meal, 1999

❑	❑	*** Fa9901 **Pig -** Stuffed.	$1.00-2.00
❑	❑	*** Fa9902 **Bunny -** Stuffed.	$1.00-2.00
❑	❑	*** Fa9903 **Chicken -** Stuffed.	$1.00-2.00
❑	❑	*** Fa9904 **Dog -** Stuffed.	$1.00-2.00
❑	❑	*** Fa9905 **Cat -** Stuffed.	$1.00-2.00
❑	❑	*** Fa9906 **Cow -** Stuffed.	$1.00-2.00

Comments: Distribution: New Zealand.

***Fa9901-06

Food Feast Happy Meal, 1999

❑ ❑ Aus Ff9901 **Cereal Bowls -** One Red & One Yellow W Characters Pictured. $2.00-4.00

❑ ❑ Aus Ff9902 **Cups -** Two Purple, Plastic Grimace Cups. $2.00-4.00

❑ ❑ Aus Ff9903 **Sipper Cup -** Ronald Pctures. $2.00-3.00

❑ ❑ Aus Ff9904 **Magnet -** Hamburglar Pictures. $2.00-3.00

❑ ❑ Aus Ff9905 **Fork & Spoon Set -** Ronald & Birdie. $1.00-2.00

Comments: Distribution: Australia: February - 1999.

Food Foolers Happy Meal, 1999

❑ ❑ USA Fo9901 **Happy Meal Box -** Turns Into Computer W 2 Screen Cards/3p. $3.00-4.00

❑ ❑ USA Fo9902 **Drink Cup -** Turns Into Signaler/Spy Toy. $3.00-4.00

❑ ❑ USA Fo9903 **McNuggets Box -** Turns Into Camera. $3.00-4.00

❑ ❑ USA Fo9904 **Fry Box -** Turns Into Cell Phone. $3.00-4.00

Comments: Distribution: USA - June 7-17, 1999. Display with the four toys was distributed to the stores, as backup following the Teenie Beanie Babies promotion.

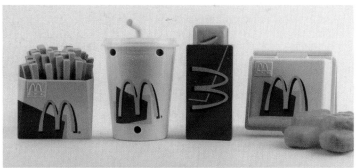

USA Fo9901-04

Furby Happy Meal, 1999

Style 1: Pushing Tail Makes Furby Growl and Moves Eyes

❑ ❑ USA Fr9901 **#1 Black W Pink Tuft & Belly.** $4.00-5.00
❑ ❑ USA Fr9902 **#2 Black W White Tuft & Belly.** $4.00-5.00
❑ ❑ USA Fr9903 **#3 Blue W Pink Tuft & Belly.** $4.00-5.00
❑ ❑ USA Fr9904 **#4 Blue W White Tuft & Belly.** $4.00-5.00
❑ ❑ USA Fr9905 **#5 Lt Green W Pink Tuft & Belly.** $4.00-5.00

❑ ❑ USA Fr9906 **#6 Lt Green W White Tuft & Belly.** $4.00-5.00

❑ ❑ USA Fr9907 **#7 Purple W Pink Tuft & Belly.** $4.00-5.00
❑ ❑ USA Fr9908 **#8 Purple W White Tuft & Belly.** $4.00-5.00
❑ ❑ USA Fr9909 **#9 Turquoise W Pink Tuft & Belly.** $4.00-5.00

❑ ❑ USA Fr9910 **#10 Turquoise W White Tuft & Belly.** $4.00-5.00

Style 2: Pushing Forward Makes Feet and Ears Move

❑ ❑ USA Fr9911 **#11 Blue W White Tuft & Belly.** $4.00-5.00
❑ ❑ USA Fr9912 **#12 Blue W Yellow Tuft & White Belly.** $4.00-5.00
❑ ❑ USA Fr9913 **#13 Gray W White Tuft & Belly.** $4.00-5.00
❑ ❑ USA Fr9914 **#14 Gray W Yellow Tuft & Belly.** $4.00-5.00
❑ ❑ USA Fr9915 **#15 Green W White Tuft & Belly.** $4.00-5.00

❑ ❑ USA Fr9916 **#16 Green W Yellow Tuft & Belly.** $4.00-5.00

❑ ❑ USA Fr9917 **#17 Orange W White Tuft & Belly.** $4.00-5.00

❑ ❑ USA Fr9918 **#18 Orange W Yellow Tuft & Belly.** $4.00-5.00

❑ ❑ USA Fr9919 **#19 Purple W White Tuft & Belly.** $4.00-5.00

❑ ❑ USA Fr9920 **#20 Purple W Yellow Tuft & Belly.** $4.00-5.00

Style 3: Pullback Furby Propels it Forward

❑ ❑ USA Fr9921 **#21 Lt Blue W Green Tuft.** $4.00-5.00
❑ ❑ USA Fr9922 **#22 Lt Blue W Purple Tuft.** $4.00-5.00
❑ ❑ USA Fr9923 **#23 Orange W Green Tuft.** $4.00-5.00
❑ ❑ USA Fr9924 **#24 Orange W Purple Tuft.** $4.00-5.00
❑ ❑ USA Fr9925 **#25 Pink W Green Tuft.** $4.00-5.00
❑ ❑ USA Fr9926 **#26 Pink W Purple Tuft.** $4.00-5.00
❑ ❑ USA Fr9927 **#27 Rust W Green Tuft.** $4.00-5.00
❑ ❑ USA Fr9928 **#28 Rust W Purple Tuft.** $4.00-5.00
❑ ❑ USA Fr9929 **#29 Yellow W Green Tuft.** $4.00-5.00
❑ ❑ USA Fr9930 **#30 Yellow W Purple Tuft.** $4.00-5.00

Fr9911-20

Fr9921-30

Fr9901-10

Style 4: Pressing Feet Makes Beak, Ears, Eyes Move
❏ ❏ USA Fr9931 **#31 Beige W Dots, White Tuft & Belly.**
$4.00-5.00
❏ ❏ USA Fr9932 **#32 Beige No Dots, Red Tuft & Belly.**
$4.00-5.00
❏ ❏ USA Fr9933 **#33 Gray W Dots, White Tuft & Belly.**
$4.00-5.00
❏ ❏ USA Fr9934 **#34 Gray No Dots, Red Tuft & White Belly.**
$4.00-5.00
❏ ❏ USA Fr9935 **#35 Orange W Dots, White Tuft & Belly.**
$4.00-5.00
❏ ❏ USA Fr9936 **#36 Orange No Dots, Red Tuft & Belly.**
$4.00-5.00
❏ ❏ USA Fr9937 **#37 Purple W Dots, White Tuft & Belly.**
$4.00-5.00
❏ ❏ USA Fr9938 **#38 Purple No Dots, Red Tuft & Belly.**
$4.00-5.00
❏ ❏ USA Fr9939 **#39 Yellow W Dots, White Tuft & Belly.**
$4.00-5.00
❏ ❏ USA Fr9940 **#40 Yellow No Dots, Red Tuft & Belly.**
$4.00-5.00

Style 5: Furby Says "Eek!" When Turned Upside Down
❏ ❏ USA Fr9941 **#41 Blue W White Tuft.** $4.00-5.00
❏ ❏ USA Fr9942 **#42 Blue W Yellow Tuft.** $4.00-5.00
❏ ❏ USA Fr9943 **#43 Lt Green W White Tuft.** $4.00-5.00
❏ ❏ USA Fr9944 **#44 Lt Green W Yellow Tuft.** $4.00-5.00
❏ ❏ USA Fr9945 **#45 Purple W White Tuft.** $4.00-5.00
❏ ❏ USA Fr9946 **#46 Purple W Yellow Tuft.** $4.00-5.00
❏ ❏ USA Fr9947 **#47 Red W White Tuft.** $4.00-5.00
❏ ❏ USA Fr9948 **#48 Red W Yellow Tuft.** $4.00-5.00
❏ ❏ USA Fr9949 **#49 White W White Tuft.** $4.00-5.00
❏ ❏ USA Fr9950 **#50 White W Yellow Tuft.** $4.00-5.00

Style 6: Furby Plays "Peekaboo" W Ears When Tail is Pressed
❏ ❏ USA Fr9951 **#51 Black W Black Tuft.** $4.00-5.00
❏ ❏ USA Fr9952 **#52 Black W Purple Tuft.** $4.00-5.00
❏ ❏ USA Fr9953 **#53 Blue W Black Tuft.** $4.00-5.00
❏ ❏ USA Fr9954 **#54 Blue W Purple Tuft.** $4.00-5.00
❏ ❏ USA Fr9955 **#55 Lt Green W Black Tuft.** $4.00-5.00
❏ ❏ USA Fr9956 **#56 Lt Green W Purple Tuft.** $4.00-5.00
❏ ❏ USA Fr9957 **#57 Orange W Black Tuft.** $4.00-5.00
❏ ❏ USA Fr9958 **#58 Orange W Purple Tuft.** $4.00-5.00
❏ ❏ USA Fr9959 **#59 Teal W Black Tuft.** $4.00-5.00
❏ ❏ USA Fr9960 **#60 Teal W Purple Tuft.** $4.00-5.00

Style 7: Pushing Tail Makes Furby Squeak and Moves Eyelids and Beak
❏ ❏ USA Fr9961 **#61 Gray W Black Tuft & Belly.** $4.00-5.00
❏ ❏ USA Fr9962 **#62 Gray W White Tuft & Belly.** $4.00-5.00
❏ ❏ USA Fr9963 **#63 Green W Black Tuft & Belly.** $4.00-5.00
❏ ❏ USA Fr9964 **#64 Green W White Tuft & Belly.**
$4.00-5.00
❏ ❏ USA Fr9965 **#65 Purple W Black Tuft & Belly.** $4.00-5.00
❏ ❏ USA Fr9966 **#66 Purple W White Tuft & Belly.**
$4.00-5.00
❏ ❏ USA Fr9967 **#67 Red W Black Tuft & Belly.** $4.00-5.00
❏ ❏ USA Fr9968 **#68 Red W White Tuft & Belly.** $4.00-5.00
❏ ❏ USA Fr9969 **#69 Yellow W Black Tuft & Belly.** $4.00-5.00
❏ ❏ USA Fr9970 **#70 Yellow W White Tuft & Belly.**
$4.00-5.00

Fr9951-60

Fr9931-40

Fr9961-70

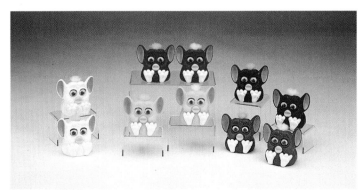

Fr9941-50

1999

Style 8: Rolling Ball on Back Moves Eyes and Ears

❏ ❏ USA Fr9971 **#71 Black W Blue Tuft & Red Belly.**
$4.00-5.00

❏ ❏ USA Fr9972 **#72 Black W Pink Tuft & Belly.** $4.00-5.00

❏ ❏ USA Fr9973 **#73 Dark Blue W Blue Tuft & Belly.**
$4.00-5.00

❏ ❏ USA Fr9974 **#74 Dark Blue W Pink Tuft & Belly.**
$4.00-5.00

❏ ❏ USA Fr9975 **#75 Light Blue W Blue Tuft & Red Belly.**
$4.00-5.00

❏ ❏ USA Fr9976 **#76 Light Blue W Pink Tuft & Belly.**
$4.00-5.00

❏ ❏ USA Fr9977 **#77 Gray W Blue Tuft & Red Belly.**
$4.00-5.00

❏ ❏ USA Fr9978 **#78 Gray W Pink Tuft & Belly.** $4.00-5.00

❏ ❏ USA Fr9979 **#79 Pink W Blue Tuft & Red Belly.**
$4.00-5.00

❏ ❏ USA Fr9980 **#80 Pink W Pink Tuft & Belly.** $4.00-5.00

Comments: Distribution: USA - 1999.

Fr9971-80

USA In9901-08

Inspector Gadget Happy Meal, 1999

❏ ❏ USA In9901 **Narvik 7 Sparker -** Chest W Head.
$4.00-5.00

❏ ❏ USA In9902 **Arm Grabber -** Left Arm. $4.00-5.00

❏ ❏ USA In9903 **Watch Belt -** Hip Area W Yellow Disc.
$4.00-5.00

❏ ❏ USA In9904 **Leg Tool -** Left Leg. $4.00-5.00

❏ ❏ USA In9905 **Leg Circuit Signaler -** Right Leg. $4.00-5.00

❏ ❏ USA In9906 **Arm Squirter -** Right Arm W Red Squirt Gun.
$4.00-5.00

❏ ❏ USA In9907 **Secret Communicator -** Gray Chest Plate.
$4.00-5.00

❏ ❏ USA In9908 **Siren Hat -** Brown W Blue Propeller.
$4.00-5.00

Comments: Distribution: USA - July 16-August 12, 1999. McDonald's latest innovation is the "Collect and Build" program. Like the French "Build a Mickey," Build an Inspector Gadget toys are based on key tools used by Inspector Gadget in the latest film. A 15" Inspector Gadget figure can be assembled with the eight toys. The Narvik 7 Sparker produces sparks inside the clear enclosed plastic chest when the spring loaded head is push down. The Arm Grabber hand opens and closes and can pick up paper. The Watch Belt is a functional digital watch with time and date features. The Leg Tool is a utility tool with pliers that open and close. The Leg Circuit Signaler is an LED signaler light. The Arm Squirter slides the hand back and forth to squirt water from the nozzle. The Trench Coat Communicator produces eight sound effects and the Siren Hat produces a mechanical siren noise when propeller is rotated. These toys are the beginning of McDonald's Collect and Build program.

McSurprise Happy Meal, 1999

❏ ❏ Hol Sp9901 **Bird -** Changes Into Ronald. $4.00-5.00

❏ ❏ Hol Sp9902 **Tiger -** Changes Into Hamburglar. $4.00-5.00

❏ ❏ Hol Sp9903 **Penguin -** Changes Into Grimace. $4.00-5.00

❏ ❏ Hol Sp9904 **Polar Bear -** Changes Into Birdie. $4.00-5.00

Comments: Regional Distribution: Holland - June, 1999. Promotion features four animals which transform into the McDonald's characters.

MscSurprise Rides Happy Meal, 1999

- ❏ ❏ Jpn Su9901 **Ronald's Delivery Truck.** $4.00-5.00
- ❏ ❏ Jpn Su9902 **Grimace's Flashing Firetruck.** $4.00-5.00
- ❏ ❏ Jpn Su9903 **Birdie's Locomotive.** $4.00-5.00
- ❏ ❏ Jpn Su9904 **Hamburglar's Flip N' Go Kart.** $4.00-5.00

Comments: Regional Distribution: Japan - February ,1999

McWave Party Happy Meal, 1999

- ❏ ❏ Zea Ex9901 **Hamburglar Snorkling -** W Blue Flippers & Cap, Arms Move, Windup. $4.00-5.00
- ❏ ❏ Zea Ex9902 **Birdie in Pink Innertube -** W Yellow Fins, Rides On Tube, Windup. $4.00-5.00
- ❏ ❏ Zea Ex9903 **Grimace Floating -** W Red Flippers, Paddles Feet, Windup. $4.00-5.00
- ❏ ❏ Zea Ex9904 **Ronald Floating -** W Red Life Jacket, Arms Move, Windup. $4.00-5.00

Comments: Distribution: New Zealand -1999; Hong Kong - June 1999.

Zea Ex9901-04

Mickey's Mystery Happy Meal, 1999

- ❏ ❏ Fra My9901 **Pen Box With Pens.** $3.00-5.00
- ❏ ❏ Fra My9902 **Compass Watch.** $3.00-5.00
- ❏ ❏ Fra My9903 **Phone Periscope.** $3.00-5.00
- ❏ ❏ Fra My9904 **Map.** $3.00-5.00

Comments: Distribution: France - June 1999.

Mister Men and Little Miss Happy Meal, 1999

- ❏ ❏ Uk Mr9901 **Miss Bossy -** Blue W Red Bow. $3.00-4.00
- ❏ ❏ Uk Mr9902 **Miss Chatterbox -** Pink W Yellow Bow, Blue Shoes. $3.00-4.00
- ❏ ❏ Uk Mr9903 **Miss Fun -** Orange W Blue Bow. $3.00-4.00
- ❏ ❏ Uk Mr9904 **Miss Giggles -** Blue W Red Hair, Yellow Nose. $3.00-4.00
- ❏ ❏ Uk Mr9905 **Miss Lucky -** Pink W Red Bow. $3.00-4.00
- ❏ ❏ Uk Mr9906 **Miss Naughty -** Purple W Green Bow. $3.00-4.00
- ❏ ❏ Uk Mr9907 **Miss Shy -** Light Blue W Pink Circle On End Of Smile. $3.00-4.00
- ❏ ❏ Uk Mr9908 **Miss Sunshine -** Yellow W Yellow Pigtails. $3.00-4.00
- ❏ ❏ Uk Mr9909 **Miss Tiny -** Pink W Blue Bow. $3.00-4.00
- ❏ ❏ Uk Mr9910 **Miss Trouble -** Yellow W Red Bow. $3.00-4.00
- ❏ ❏ Uk Mr9911 **Mr. Bounce -** Yellow W Pink Hat. $3.00-4.00
- ❏ ❏ Uk Mr9912 **Mr. Bump -** Light Blue. $3.00-4.00
- ❏ ❏ Uk Mr9913 **Mr. Cheerful -** Orange W Yellow Hat W Blue Band On Hat. $3.00-4.00
- ❏ ❏ Uk Mr9914 **Mr. Clever -** Orange W Green Hat. $3.00-4.00
- ❏ ❏ Uk Mr9915 **Mr. Chatterbox -** Red W Green Hat. $3.00-4.00
- ❏ ❏ Uk Mr9916 **Mr. Clumsy -** Green W Yellow Nose And Darker Green Hair. $3.00-4.00
- ❏ ❏ Uk Mr9917 **Mr. Worry -** Blue W Red Nose. $3.00-4.00
- ❏ ❏ Uk Mr9918 **Mr. Forgetful -** Purple W Red Hat. $3.00-4.00
- ❏ ❏ Uk Mr9919 **Mr. Funny -** Lime Green W Yellow Hat W Blue Hat Band. $3.00-4.00
- ❏ ❏ Uk Mr9920 **Mr. Fussy -** Green W Yellow Nose. $3.00-4.00
- ❏ ❏ Uk Mr9921 **Mr. Nonsense -** Yellow W Green Hat. $3.00-4.00
- ❏ ❏ Uk Mr9922 **Mr. Greedy -** Neon Pink W Arms Raised Up. $3.00-4.00
- ❏ ❏ Uk Mr9923 **Mr. Grumble -** Purple W Light Blue Hat. $3.00-4.00
- ❏ ❏ Uk Mr9924 **Mr. Grumpy -** Light Blue W Green Hat & Frown On Face. $3.00-4.00
- ❏ ❏ Uk Mr9925 **Mr. Happy -** Yellow W Smile. $3.00-4.00
- ❏ ❏ Uk Mr9926 **Mr. Impossible -** Purple W Blue Hat. $3.00-4.00
- ❏ ❏ Uk Mr9927 **Mr. Jelly -** Red/Pinkish W Arms At Side. $3.00-4.00
- ❏ ❏ Uk Mr9928 **Mr. Lazy -** Yellow W Blue Hat. $3.00-4.00
- ❏ ❏ Uk Mr9929 **Mr. Mean -** Light Blue W Yellow Nose. $3.00-4.00
- ❏ ❏ Uk Mr9930 **Mr.Muddle -** Light Green W Red Hat. $3.00-4.00
- ❏ ❏ Uk Mr9931 **Mr. Mischief -** Yellow W Red/Pinkish Hat. $3.00-4.00
- ❏ ❏ Uk Mr9932 **Mr. Noisy -** Red W Large Black Mouth. $3.00-4.00
- ❏ ❏ Uk Mr9933 **Mr. Nosey -** Green W Large Smile. $3.00-4.00
- ❏ ❏ Uk Mr9934 **Mr. Rush -** Purple W Yellow Hat. $3.00-4.00
- ❏ ❏ Uk Mr9935 **Mr. Silly -** Yellow W Orange Hat. $3.00-4.00
- ❏ ❏ Uk Mr9936 **Mr. Small -** Red W Blue Hat. $3.00-4.00

1999

		Uk Mr9937 **Mr. Sneeze -** Blue W Red Nose.	$3.00-4.00
		Uk Mr9938 **Mr. Strong -** Red W Green Hat.	$3.00-4.00
		Uk Mr9939 **Mr. Tall -** Blue W Red Feet.	$3.00-4.00
		Uk Mr9940 **Mr. Uppity -** Red W Black Hat.	$3.00-4.00

Comments: Distribution: UK - June, 1999.

Mulan Happy Meal, 1999

| | | USA Mu9901 **Ling -** On Red Base Spinner. | $2.00-3.00 |

USA Mu9902 **Mulan & Cri-Kee -** Wearing Green Shirt Holding Cri-Kee On Her Back/2p Launcher. $2.00-3.00

USA Mu9903 **Khan -** Wearing Black Cape Spinner. $2.00-3.00

USA Mu9904 **Mushu -** Gray Statue W Red Figure On Disc/2p Launcher. $2.00-3.00

USA Mu9905 **Yao -** Wearing Red Trim Spinner. $2.00-3.00

USA Mu9906 **Shang -** Wearing Green Trim/2p Launcher. $2.00-3.00

USA Mu9907 **Chien Po -** Wearing Grey W Blue Trim Spinner. $2.00-3.00

USA Mu9908 **Shan Yu -** Wearing Blue W Brown Bird/2p Launcher. $2.00-3.00

Comments: Distribution: USA - 1999.

USA Mu9926

USA Mu9901-04

USA Mu9905-08

USA Mu9945

Mystic Knights Happy Meal, 1999

❏ ❏ USA Mk9901 **#1 Rohan -** Gold Knight W Red Shield.
$2.00-3.00

❏ ❏ USA Mk9902 **#2 Queen Maeve -** Gray Long Dress, Brown Hair & Maroon Cloth Cape. $2.00-3.00

❏ ❏ USA Mk9903 **#3 Angus -** Blue Knight W Gold Accents & Green Shield. $2.00-3.00

❏ ❏ USA Mk9904 **#4 Corc -** Brown Knight W Gray Cape/2p.
$2.00-3.00

❏ ❏ USA Mk9905 **#5 Deirdre -** White Short Kilted Knight W Turquoise/Gold Shield/2p.. $2.00-3.00

❏ ❏ USA Mk9906 **#6 Mider -** Tan Body W Dark Brown Long Coat. $2.00-3.00

❏ ❏ USA Mk9907 **#7 Ivar -** Blue Knight W Silver Shield/2p.
$2.00-3.00

❏ ❏ USA Mk9908 **#8 Lugad -** Black & Gold Husky Fig W White Masked Face & Red Hat. $2.00-3.00

❏ ❏ USA Mk9909 **#9 Dragon -** 4 Extra Pieces from Set #2, 4, 6 & 8. $5.00-8.00

Comments: National Distribution: USA, Canada - April, 1999. Dragon, USA Re9909 is formed by combining an extra piece from packages #2, 4, 6, and 8 together.

USA Mk9901-08

USA Mk9909. Dragon.

USA Mk9927. Happy Meal Color Card.

Olympic Games Parade Happy Meal, 1999

☐ ☐	Aus OI9901	**Birdie Basketball.**	$4.00-5.00
☐ ☐	Aus OI9902	**Ronald Discus Throwing.**	$4.00-5.00
☐ ☐	Aus OI9903	**Grimace Weight Lifting.**	$4.00-5.00
☐ ☐	Aus OI9904	**Hamburglar Kayaking.**	

Comments: Distribution: Australia - May, 1999. The four pieces snap together at the base, like the Space Jam set.

Reversible Vehicles Happy Meal, 1999

☐ ☐	Zea Re9901	**Train Engine -** Green/Gray Changes To Orange/Gray.	$4.00-5.00
☐ ☐	Zea Re9902	**Car -** Blue Changes To Yellow/Blue.	$4.00-5.00
☐ ☐	Zea Re9903	**Fire Truck -** Red/Yellow Changes To Yellow/Red.	$4.00-5.00
☐ ☐	Zea Re9904	**Car -** Purple/Pink Changes To Green/Pink.	$4.00-5.00

Comments: Distribution - New Zealand, 1999. Each of the vehicles reverses the top of the cab color by switching the door panels, the bottom panel, and the cab panel.

Zea Re9901-04

Zea Re9901-04

Snoopy World Tour Happy Meal, 1999

☐ ☐	Jpn Sn9901	**# I Snoopy in Taiwan -** Red Jacket .	$6.00-8.00
☐ ☐	Jpn Sn9902	**# 2 Snoopy in Scotland -** Grey Hat & Shirt.	$6.00-8.00
☐ ☐	Jpn Sn9903	**# 3 Snoopy in Texas Wild West/USA -** Red Cowboy Hat.	$6.00-8.00
☐ ☐	Jpn Sn9904	**# 4 Snoopy in Malaysia -** Green Shirt W Flower.	$6.00-8.00

USA Sn9901-04

☐ ☐	Jpn Sn9905	**# 5 Snoopy in Germany -** Blue Hat.	$6.00-8.00
☐ ☐	Jpn Sn9906	**# 6 Snoopy in the U.S.A. -** Red, White & Blue Hat.	$6.00-8.00
☐ ☐	Jpn Sn9907	**# 7 Snoopy in the United Kingdom -** Black Hat.	$6.00-8.00
☐ ☐	Jpn Sn9908	**# 8 Snoopy in China -** Red Suit & Black Hat.	$6.00-8.00

USA Sn9905-08

Jpn Sn9909 # 9 **Snoopy in Indonesia** - Red Hat W Racket.
$6.00-8.00

Jpn Sn9910 #10 **Snoopy in Switzerland** - Black, Yellow & White Vest. $6.00-8.00

Jpn Sn9911 #11 **Snoopy in Singapore** - Blue Shirt.
$6.00-8.00

Jpn Sn9912 #12 **Snoopy in Japan** - Black Shirt.
$6.00-8.00

Jpn Sn9913 #13 **Snoopy in the Phillippines** - Red/Yellow Hat & Shirt. $6.00-8.00

Jpn Sn9914 #14 **Snoopy in Italy** - Blue/Red Hat & Shirt/2p.
$6.00-8.00

Jpn Sn9915 #15 **Snoopy in Latin America** - Blue & Yellow Shirt. $6.00-8.00

Jpn Sn9916 #16 **Snoopy in Alaska/USA** - Green Hat & Coat. $6.00-8.00

USA Sn9913-16

Jpn Sn9917 #17 **Snoopy in France** - Lavender Shirt.
$6.00-8.00

Jpn Sn9918 #18 **Snoopy in Australia** - Brown/Yellow Hat & Shirt. $6.00-8.00

Jpn Sn9919 #19 **Snoopy in Korea** - Blue & Red Hat.
$6.00-8.00

Jpn Sn9920 #20 **Snoopy in Mexico** - Blue, Red, Green & Yellow Hat/2p. $6.00-8.00

USA Sn9917-20

Jpn Sn9921 #21 **Snoopy in Hong Kong** - Yellow Hat & Shirt. $6.00-8.00

Jpn Sn9922 #22 **Snoopy in Spain** - Red Hat & Shirt.
$6.00-8.00

Jpn Sn9923 #23 **Snoopy in New Zealand** - Red Hat & Shirt W Yellow Paddle. $6.00-8.00

Jpn Sn9924 #24 **Snoopy in Thailand** - No Hat W White Shirt. $6.00-8.00

Jpn Sn9925 #25 **Snoopy in Norway** - Yellow Horns On Hat. $6.00-8.00

Jpn Sn9926 #26 **Snoopy in Canada** - Brown & Red Hat/2p.
$6.00-8.00

Jpn Sn9927 #27 **Snoopy in Hawaii/USA** - Purple Sunglasses & Shirt. $6.00-8.00

Jpn Sn9928 #28 **Snoopy in Mongolia** - Brown & White Hat. $6.00-8.00

USA Sn9921-24

*** Sn9929 #29 **Snoopy in Peru** - Red Multicolored Cape/ Hat. $6.00-10.00

*** Sn9930 #30 **Snoopy in Venezuela** - Large Brown Hat/ Holding Shakers In Hands. $6.00-10.00

*** Sn9931 #31 **Snoopy in Panama** $6.00-10.00

Comments: Distribution: Japan, UK - 1999. The original set of twenty-eight Snoopy figures came with MOVEABLE arms and legs, NOT A RUBBER version. In some stores, a complete set in a blue carrying case could be obtained. As of this writing, a bogus, unauthorized set has appeared at some toy shows. The 1999 Bogus Snoopy Set has a smaller carrying case, identical to the original, but with RUBBER Snoopy figurines. This rubberized version of Snoopy was not distributed across the counter at McDonald's, according to our research in early 1999.

USA Sn9909-12

USA Sn9925-28

Original Snoopy Case.

Tarzan Happy Meal, 1999

❏ ❏ USA Ta9901 **#1 Tarzan** - Tan W Brown Long Hair And Briefs Standing On Grn Log/2 p. $2.00-3.00

❏ ❏ USA Ta9902 **#2 Terk** - Purple Monkey W Brown Barrel On Head. $2.00-3.00

❏ ❏ USA Ta9903 **#3 Jane** - Yellow Long Dress W Yellow Umbrella Top/2p. $2.00-3.00

❏ ❏ USA Ta9904 **#4 Tantor** - Maroon Elephant. $2.00-3.00

❏ ❏ USA Ta9905 **#5 Porter** - Tan Professor On Gray Tricycle. $2.00-3.00

❏ ❏ USA Ta9906 **#6 Kala** - Brown Ape W Boy On Back. $2.00-3.00

❏ ❏ USA Ta9907 **#7 Clayton** - Man W Yel Shirt, Grn Pants, Wht Net, Gray Sword/3p. $2.00-3.00

❏ ❏ USA Ta9908 **#8 Sabor** - Yellow Leopard W Orange Eyes & Brn Spots. $2.00-3.00

Comments: National Distribution: USA, Canada - June 1999.

Original tray of figures from Japan.

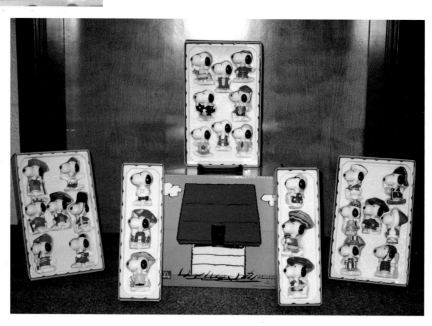

Original Carrying Case with Snoopy figures.

Teenie Beanie Babies III Happy Meal, 1999

❏ ❏ USA Ty9901 **#1 Freckles the Leopard -** Tan, Orange and Brown Spots. $3.00-5.00

❏ ❏ USA Ty9902 **#2 Antsy the Anteater -** Gray, Black and White. $3.00-4.00

❏ ❏ USA Ty9903 **#3 Smoochy the Frog -** Green With Yellow Feet and Eyes. $3.00-4.00

❏ ❏ USA Ty9904 **#4 Spunky the Cocker Spaniel -** Tan. $4.00-5.00

❏ ❏ USA Ty9905 **#5 Rocket the Blue Jay -** Blue and White Bird W Black Feet and Bill. $3.00-4.00

❏ ❏ USA Ty9906 **#6 Iggy the Iguana -** Blue/Green W Lime Green Mouth. $3.00-4.00

❏ ❏ USA Ty9907 **#7 Strut the Rooster -** Pink/Reddish W Red Tail and Yellow Feet. $3.00-4.00

❏ ❏ USA Ty9908 **#8 Nuts the Squirrel -** Tan With White Belly, Standing. $3.00-4.00

USA Ty9901-04

USA Ty9905-08

❏ ❏ USA Ty9909 **#9 Claude the Crab -** Brown W Dark Accents. $3.00-4.00

❏ ❏ USA Ty9910 **#10 Stretchy the Ostrich -** Tan Head and Feet W Brown and White Body. $3.00-4.00

❏ ❏ USA Ty9911 **#11 Nooky the Husky -** Gray W White Belly and Feet. $3.00-4.00

❏ ❏ USA Ty9912 **#12 Chip the Cat -** Half Brown and Black W White Paws. $3.00-4.00

Comments: National Distribution: USA, Canada - June, 1999.

USA Ty9909-12

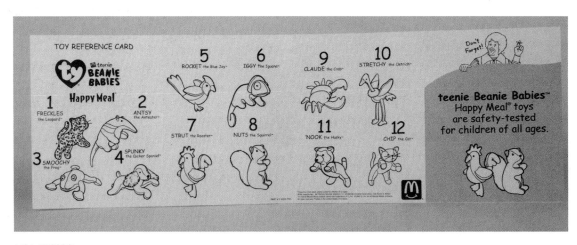

USA Ty9928